INTERNET THEORY, TECHNOLOGY AND APPLICATIONS

BIG DATA

AN EXPLORATION OF OPPORTUNITIES, VALUES, AND PRIVACY ISSUES

INTERNET THEORY, TECHNOLOGY AND APPLICATIONS

Additional books in this series can be found on Nova's website under the Series tab.

Additional e-books in this series can be found on Nova's website under the e-book tab.

INTERNET THEORY, TECHNOLOGY AND APPLICATIONS

BIG DATA

AN EXPLORATION OF OPPORTUNITIES, VALUES, AND PRIVACY ISSUES

CODY AGNELLUTTI
EDITOR

Copyright © 2014 by Nova Science Publishers, Inc.

All rights reserved. No part of this book may be reproduced, stored in a retrieval system or transmitted in any form or by any means: electronic, electrostatic, magnetic, tape, mechanical photocopying, recording or otherwise without the written permission of the Publisher.

For permission to use material from this book please contact us:
Telephone 631-231-7269; Fax 631-231-8475
Web Site: http://www.novapublishers.com

NOTICE TO THE READER

The Publisher has taken reasonable care in the preparation of this book, but makes no expressed or implied warranty of any kind and assumes no responsibility for any errors or omissions. No liability is assumed for incidental or consequential damages in connection with or arising out of information contained in this book. The Publisher shall not be liable for any special, consequential, or exemplary damages resulting, in whole or in part, from the readers' use of, or reliance upon, this material. Any parts of this book based on government reports are so indicated and copyright is claimed for those parts to the extent applicable to compilations of such works.

Independent verification should be sought for any data, advice or recommendations contained in this book. In addition, no responsibility is assumed by the publisher for any injury and/or damage to persons or property arising from any methods, products, instructions, ideas or otherwise contained in this publication.

This publication is designed to provide accurate and authoritative information with regard to the subject matter covered herein. It is sold with the clear understanding that the Publisher is not engaged in rendering legal or any other professional services. If legal or any other expert assistance is required, the services of a competent person should be sought. FROM A DECLARATION OF PARTICIPANTS JOINTLY ADOPTED BY A COMMITTEE OF THE AMERICAN BAR ASSOCIATION AND A COMMITTEE OF PUBLISHERS.

Additional color graphics may be available in the e-book version of this book.

Library of Congress Cataloging-in-Publication Data

ISBN: 978-1-63321-397-5

Published by Nova Science Publishers, Inc. † New York

CONTENTS

Preface		vii
Chapter 1	Big Data: Seizing Opportunities, Preserving Values *Executive Office of the President*	1
Chapter 2	Big Data and Privacy: A Technological Perspective *President's Council of Advisors on Science and Technology*	93
Index		169

PREFACE

Chapter 1 – We are living in the midst of a social, economic, and technological revolution. How we communicate, socialize, spend leisure time, and conduct business has moved onto the Internet. The Internet has in turn moved into our phones, into devices spreading around our homes and cities, and into the factories that power the industrial economy. The resulting explosion of data and discovery is changing our world.

Most definitions of big data reflect the growing technological ability to capture, aggregate, and process an ever-greater volume, velocity, and variety of data.

Big data will transform the way we live and work and alter the relationships between government, citizens, businesses, and consumers. This review focuses on how the public and private sectors can maximize the benefits of big data while minimizing its risks. It also identifies opportunities for big data to grow our economy, improve health and education, and make our nation safer and more energy efficient.

Chapter 2 – The ubiquity of computing and electronic communication technologies has led to the exponential growth of data from both digital and analog sources. New technical abilities to gather, analyze, disseminate, and preserve vast quantities of data raise new concerns about the nature of privacy and the means by which individual privacy might be compromised or protected.

This report begins by exploring the changing nature of privacy as computing technology has advanced and big data has come to the forefront. It proceeds by identifying the sources of these data, the utility of these data — including new data analytics enabled by data mining and data fusion — and the privacy challenges big data poses in a world where technologies for re-identification often outpace privacy-preserving de-identification capabilities, and where it is increasingly hard to identify privacy-sensitive information at the time of its collection.

In: Big Data
Editor: Cody Agnellutti

ISBN: 978-1-63321-397-5
© 2014 Nova Science Publishers, Inc.

Chapter 1

BIG DATA: SEIZING OPPORTUNITIES, PRESERVING VALUES[*]

Executive Office of the President

I. BIG DATA AND THE INDIVIDUAL

What Is Big Data?

Since the first censuses were taken and crop yields recorded in ancient times, data collection and analysis have been essential to improving the functioning of society. Foundational work in calculus, probability theory, and statistics in the 17th and 18th centuries provided an array of new tools used by scientists to more precisely predict the movements of the sun and stars and determine population-wide rates of crime, marriage, and suicide. These tools often led to stunning advances. In the 1800s, Dr. John Snow used early modern data science to map cholera "clusters" in London. By tracing to a contaminated public well a disease that was widely thought to be caused by "miasmatic" air, Snow helped lay the foundation for the germ theory of disease.[1]

Gleaning insights from data to boost economic activity also took hold in American industry. Frederick Winslow Taylor's use of a stopwatch and a clipboard to analyze productivity at Midvale Steel Works in Pennsylvania

---

[*] This is an edited, reformatted and augmented version of report released May 2014.

increased output on the shop floor and fueled his belief that data science could revolutionize every aspect of life.[2] In 1911, Taylor wrote *The Principles of Scientific Management* to answer President Theodore Roosevelt's call for increasing "national efficiency":

> [T]he fundamental principles of scientific management are applicable to all kinds of human activities, from our simplest individual acts to the work of our great corporations.... [W]henever these principles are correctly applied, results must follow which are truly astounding.[3]

Today, data is more deeply woven into the fabric of our lives than ever before. We aspire to use data to solve problems, improve well-being, and generate economic prosperity. The collection, storage, and analysis of data is on an upward and seemingly unbounded trajectory, fueled by increases in processing power, the cratering costs of computation and storage, and the growing number of sensor technologies embedded in devices of all kinds. In 2011, some estimated the amount of information created and replicated would surpass 1.8 zettabytes.[4] In 2013, estimates reached 4 zettabytes of data generated worldwide.[5]

WHAT IS A ZETTABYTE?

A zettabyte is 1,000,000,000,000,000,000,000 bytes, or units of information. Consider that a single byte equals one character of text. The 1,250 pages of Leo Tolstoy's *War and Peace* would fit into a zettabyte 323 trillion times.[6] Or imagine that every person in the United States took a digital photo every second of every day for over a month. All of those photos put together would equal about one zettabyte.

More than 500 million photos are uploaded and shared every day, along with more than 200 hours of video every minute. But the volume of information that people create themselves—the full range of communications from voice calls, emails and texts to uploaded pictures, video, and music—pales in comparison to the amount of digital information created *about* them each day.

These trends will continue. We are only in the very nascent stage of the so-called "Internet of Things," when our appliances, our vehicles and a growing set of "wearable" technologies will be able to communicate with each other. Technological advances have driven down the cost of creating, capturing, managing, and storing information to onesixth of what it was in 2005. And since 2005, business investment in hardware, software, talent, and services has increased as much as 50 percent, to $4 trillion.

> ## THE "INTERNET OF THINGS"
>
> The "Internet of Things" is a term used to describe the ability of devices to communicate with each other using embedded sensors that are linked through wired and wireless networks. These devices could include your thermostat, your car, or a pill you swallow so the doctor can monitor the health of your digestive tract. These connected devices use the Internet to transmit, compile, and analyze data.

There are many definitions of "big data" which may differ depending on whether you are a computer scientist, a financial analyst, or an entrepreneur pitching an idea to a venture capitalist. Most definitions reflect the growing technological ability to capture, aggregate, and process an ever-greater volume, velocity, and variety of data. In other words, "data is now available faster, has greater coverage and scope, and includes new types of observations and measurements that previously were not available."[7] More precisely, big datasets are "large, diverse, complex, longitudinal, and/or distributed datasets generated from instruments, sensors, Internet transactions, email, video, click streams, and/or all other digital sources available today and in the future."[8]

What really matters about big data is what it does. Aside from how we define big data as a technological phenomenon, the wide variety of potential uses for big data analytics raises crucial questions about whether our legal, ethical, and social norms are sufficient to protect privacy and other values in a big data world. Unprecedented computational power and sophistication make possible unexpected discoveries, innovations, and advancements in our quality of life. But these capabilities, most of which are not visible or available to the average consumer, also create an asymmetry of power between those who hold the data and those who intentionally or inadvertently supply it.

Part of the challenge, too, lies in understanding the many different contexts in which big data comes into play. Big data may be viewed as property, as a public resource, or as an expression of individual identity.[9] Big data applications may be the driver of America's economic future or a threat to cherished liberties. Big data may be all of these things. For the purposes of this 90-day study, the review group does not purport to have all the answers to big data. Both the technology of big data and the industries that support it are constantly innovating and changing. Instead, the study focuses on asking the most important questions about the relationship between individuals and those who collect and use data about them.

> ## THE SCOPE OF THIS REVIEW
>
> On January 17, in a speech at the Justice Department about reforming the United States' signals intelligence practices, President Obama tasked his Counselor John Podesta with leading a comprehensive review of the impact big data technologies are having, and will have, on a range of economic, social, and government activities. Podesta was joined in this effort by Secretary of Commerce Penny Pritzker, Secretary of Energy Ernest Moniz, the President's Science Advisor John Holdren, the President's Economic Advisor Jeffrey Zients, and other senior government officials. The President's Council of Advisors for Science & Technology conducted a parallel report to take measure of the underlying technologies. Their findings underpin many of the technological assertions in this report.
>
> This review was conceived as fundamentally a scoping exercise. Over 90 days, the review group engaged with academic experts, industry representatives, privacy advocates, civil rights groups, law enforcement agents, and other government agencies. The White House Office of Science and Technology Policy jointly organized three university conferences, at the Massachusetts Institute of Technology, New York University, and the University of California, Berkeley. The White House Office of Science & Technology Policy also issued a "Request for Information" seeking public comment on issues of big data and privacy and received more than 70 responses. In addition, the WhiteHouse.gov platform was used to conduct an unscientific survey of public attitudes about different uses of big data and various big data technologies. A list of the working group's activities can be found in the Appendix.

What Is Different about Big Data?

This section begins by defining what is truly new and different about big data, drawing on the work of the President's Council of Advisors on Science

& Technology (PCAST), which has worked in parallel on a separate report, "Big Data and Privacy: A Technological Perspective."[10]

The "3 Vs": Volume, Variety and Velocity

For purposes of this study, the review group focused on data that is so large in volume, so diverse in variety or moving with such velocity, that traditional modes of data capture and analysis are insufficient—characteristics colloquially referred to as the "3 Vs." The declining cost of collection, storage, and processing of data, combined with new sources of data like sensors, cameras, geospatial and other observational technologies, means that we live in a world of near-ubiquitous data collection. The volume of data collected and processed is unprecedented. This explosion of data—from web-enabled appliances, wearable technology, and advanced sensors to monitor everything from vital signs to energy use to a jogger's running speed—will drive demand for high-performance computing and push the capabilities of even the most sophisticated data management technologies.

There is not only more data, but it also comes from a wider variety of sources and formats. As described in the report by the President's Council of Advisors of Science & Technology, some data is "born digital," meaning that it is created specifically for digital use by a computer or data processing system. Examples include email, web browsing, or GPS location. Other data is "born analog," meaning that it emanates from the physical world, but increasingly can be converted into digital format. Examples of analog data include voice or visual information captured by phones, cameras or video recorders, or physical activity data, such as heart rate or perspiration monitored by wearable devices.[11] With the rising capabilities of "data fusion," which brings together disparate sources of data, big data can lead to some remarkable insights.

Furthermore, data collection and analysis is being conducted at a velocity that is increasingly approaching real time, which means there is a growing potential for big data analytics to have an immediate effect on a person's surrounding environment or decisions being made about his or her life. Examples of high-velocity data include click-stream data that records users' online activities as they interact with web pages, GPS data from mobile devices that tracks location in real time, and social media that is shared broadly. Customers and companies are increasingly demanding that this data be analyzed to benefit them instantly. Indeed, a mobile mapping application is essentially useless if it cannot immediately and accurately identify the phone's location, and real-time processing is critical in the computer systems that ensure the safe operation of our cars.

> ## WHAT ARE THE SOURCES OF BIG DATA?
>
> The sources and formats of data continue to grow in variety and complexity. A partial list of sources includes the public web; social media; mobile applications; federal, state and local records and databases; commercial databases that aggregate individual data from a spectrum of commercial transactions and public records; geospatial data; surveys; and traditional offline documents scanned by optical character recognition into electronic form. The advent of the more Internet-enabled devices and sensors expands the capacity to collect data from physical entities, including sensors and radio-frequency identification (RFID) chips. Personal location data can come from GPS chips, cell-tower triangulation of mobile devices, mapping of wireless networks, and in-person payments.[12]

New Opportunities, New Challenges

Big data technologies can derive value from large datasets in ways that were previously impossible—indeed, big data can generate insights that researchers didn't even think to seek. But the technical capabilities of big data have reached a level of sophistication and pervasiveness that demands consideration of how best to balance the opportunities afforded by big data against the social and ethical questions these technologies raise.

The Power and Opportunity of Big Data Applications

Used well, big data analysis can boost economic productivity, drive improved consumer and government services, thwart terrorists, and save lives. Examples include:

- Big data and the growing "Internet of Things" have made it possible to merge the industrial and information economies. Jet engines and delivery trucks can now be outfitted with sensors that monitor hundreds of data points and send automatic alerts when maintenance is needed.[13] This makes repairs smoother, reducing maintenance costs and increasing safety.

- The Centers for Medicare and Medicaid Services have begun using predictive analytics software to flag likely instances of reimbursement fraud before claims are paid. The Fraud Prevention System helps identify the highest risk health care providers for fraud, waste and abuse in real time, and has already stopped, prevented or identified $115 million in fraudulent payments—saving $3 for every $1 spent in the program's first year.[14]
- During the most violent years of the war in Afghanistan, the Defense Advanced Research Projects Agency (DARPA) deployed teams of data scientists and visualizers to the battlefield. In a program called Nexus 7, these teams embedded directly with military units and used their tools to help commanders solve specific operational challenges. In one area, Nexus 7 engineers fused satellite and surveillance data to visualize how traffic flowed through road networks, making it easier to locate and destroy improvised explosive devices.
- One big data study synthesized millions of data samples from monitors in a neonatal intensive care unit to determine which newborns were likely to contract potentially fatal infections. By analyzing all of the data—not just what doctors noted on their rounds—the project was able to identify factors, like increases in temperature and heart rate, that serve as early warning signs that an infection may be taking root. These early signs of infection are not something even an experienced and attentive doctor would catch through traditional practices.[15]

Big data technology also holds tremendous promise for better managing demand across electricity grids, improving energy efficiency, boosting agricultural productivity in the developing world, and projecting the spread of infectious diseases, among other applications.

Finding the Needle in the Haystack

Computational capabilities now make "finding a needle in a haystack" not only possible, but practical. In the past, searching large datasets required both rationally organized data and a specific research question, relying on choosing the right query to return the correct result. Big data analytics enable data scientists to amass lots of data, including unstructured data, and find anomalies or patterns. A key privacy challenge in this model of discovery is that in order to find the needle, you have to have a haystack. To obtain certain insights, you need a certain quantity of data.

For example, a genetic researcher at the Broad Institute found that having a large number of genetic datasets makes the critical difference in identifying the meaningful genetic variant for a disease. In this research, a genetic variant related to schizophrenia was not detectable when analyzed in 3,500 cases, and was only weakly identifiable using 10,000 cases, but was suddenly statistically significant with 35,000 cases. As the researcher observed, "There is an inflection point at which everything changes."[16] The need for vast quantities of data—particularly personally sensitive data like genetic data—is a significant challenge for researchers for a variety of reasons, but notably because of privacy laws that limit access to data.

The data clusters and relationships revealed in large data sets can be unexpected but deliver incisive results. On the other hand, even with lots of data, the information revealed by big data analysis isn't necessarily perfect. Identifying a pattern doesn't establish whether that pattern is significant. Correlation still doesn't equal causation. Finding a correlation with big data techniques may not be an appropriate basis for predicting outcomes or behavior, or rendering judgments on individuals. In big data, as with all data, interpretation is always important.

The Benefits and Consequences of Perfect Personalization

The fusion of many different kinds of data, processed in real time, has the power to deliver exactly the right message, product, or service to consumers before they even ask. Small bits of data can be brought together to create a clear picture of a person to predict preferences or behaviors. These detailed personal profiles and personalized experiences are effective in the consumer marketplace and can deliver products and offers to precise segments of the population—like a professional accountant with a passion for knitting, or a home chef with a penchant for horror films.

Unfortunately, "perfect personalization" also leaves room for subtle and not-so-subtle forms of discrimination in pricing, services, and opportunities. For example, one study found web searches involving black-identifying names (e.g., "Jermaine") were more likely to display ads with the word "arrest" in them than searches with white-identifying names (e.g., "Geoffrey"). This research was not able determine exactly why a racially biased result occurred, recognizing that ad display is algorithmically generated based on a number of variables and decision processes.[17] But it's clear that outcomes like these, by serving up different kinds of information to different groups, have the potential to cause real harm to individuals, whether they are pursuing a job, purchasing a home, or simply searching for information.

Another concern is that big data technology could assign people to ideologically or culturally segregated enclaves known as "filter bubbles" that effectively prevent them from encountering information that challenges their biases or assumptions.[18] Extensive profiles about individuals and their preferences are being painstakingly developed by companies that acquire and process increasing amounts of data. Public awareness of the scope and scale of these activities is limited, however, and consumers have few opportunities to control the collection, use, and re-use of these data profiles.

De-Identification and Re-Identification
As techniques like data fusion make big data analytics more powerful, the challenges to current expectations of privacy grow more serious. When data is initially linked to an individual or device, some privacy-protective technology seeks to remove this linkage, or "de-identify" personally identifiable information—but equally effective techniques exist to pull the pieces back together through "re-identification." Similarly, integrating diverse data can lead to what some analysts call the "mosaic effect," whereby personally identifiable information can be derived or inferred from datasets that do not even include personal identifiers, bringing into focus a picture of who an individual is and what he or she likes.

Many technologists are of the view that de-identification of data as a means of protecting individual privacy is, at best, a limited proposition.[19] In practice, data collected and deidentified is protected in this form by companies' commitments to not re-identify the data and by security measures put in place to ensure those protections. Encrypting data, removing unique identifiers, perturbing data so it no longer identifies individuals, or giving users more say over how their data is used through personal profiles or controls are some of the current technological solutions. But meaningful de-identification may strip the data of both its usefulness and the ability to ensure its provenance and accountability. Moreover, it is difficult to predict how technologies to re-identify seemingly anonymized data may evolve. This creates substantial uncertainty about how an individual controls his or her own information and identity, and how he or she disputes decisionmaking based on data derived from multiple datasets.

The Persistence of Data
In the past, retaining physical control over one's personal information was often sufficient to ensure privacy. Documents could be destroyed, conversations forgotten, and records expunged. But in the digital world,

information can be captured, copied, shared, and transferred at high fidelity and retained indefinitely. Volumes of data that were once unthinkably expensive to preserve are now easy and affordable to store on a chip the size of a grain of rice. As a consequence, data, once created, is in many cases effectively permanent. Furthermore, digital data often concerns multiple people, making personal control impractical. For example, who owns a photo—the photographer, the people represented in the image, the person who first posted it, or the site to which it was posted? The spread of these new technologies are fundamentally changing the relationship between a person and the data about him or her.

Certainly data is freely shared and duplicated more than ever before. The specific responsibilities of individuals, government, corporations, and the network of friends, partners, and other third parties who may come into possession of personal data have yet to be worked out. The technological trajectory, however, is clear: more and more data will be generated about individuals and will persist under the control of others. Ensuring that data is secure is a matter of the utmost importance. For that reason, models for public-private cooperation, like the Administration's Cybersecurity Framework, launched in February 2014, are a critical part of ensuring the security and resiliency of the critical infrastructure supporting much of the world's data assets.[20]

Affirming Our Values

No matter how serious and consequential the questions posed by big data, this Administration remains committed to supporting the digital economy and the free flow of data that drives its innovation. The march of technology always raises questions about how to adapt our privacy and social values in response. The United States has met this challenge through considered debate in the public sphere, in the halls of Congress, and in the courts—and throughout its history has consistently been able to realize the rights enshrined in the Constitution, even as technology changes.

Since the earliest days of President Obama's first term, this Administration has called on both the public and private sector to harness the power of data in ways that boost productivity, improve lives, and serve communities. That said, this study is about more than the capabilities of big data technologies. It is also about how big data may challenge fundamental American values and existing legal frameworks. This report focuses on the federal government's role in assuring that our values endure and our laws evolve as big data technologies change the landscape for consumers and citizens.

In the last year, the public debate on privacy has largely focused on how government, particularly the intelligence community, collects, stores, and uses data. This report largely leaves issues raised by the use of big data in signals intelligence to be addressed through the policy guidance that the President announced in January. However, this report considers many of the other ways government collects and uses large datasets for the public good. Public trust is required for the proper functioning of government, and governments must be held to a higher standard for the collection and use of personal data than private actors. As President Obama has unequivocally stated, "It is not enough for leaders to say: trust us, we won't abuse the data we collect."[21]

Recognizing that big data technologies are used far beyond the intelligence community, this report has taken a broad view of the issues implicated by big data. These new technologies do not only test individual privacy, whether defined as the right to be let alone, the right to control one's identity, or some other variation. Some of the most profound challenges revealed during this review concern how big data analytics may lead to disparate inequitable treatment, particularly of disadvantaged groups, or create such an opaque decision-making environment that individual autonomy is lost in an impenetrable set of algorithms.

These are not unsolvable problems, but they merit deep and serious consideration. The historian Melvin Kranzberg's First Law of Technology is important to keep in mind: "Technology is neither good nor bad; nor is it neutral."[22] Technology can be used for the public good, but so too can it be used for individual harm. Regardless of technological advances, the American public retains the power to structure the policies and laws that govern the use of new technologies in a way that protects foundational values.

Big data is changing the world. But it is not changing Americans' belief in the value of protecting personal privacy, of ensuring fairness, or of preventing discrimination. This report aims to encourage the use of data to advance social good, particularly where markets and existing institutions do not otherwise support such progress, while at the same time supporting frameworks, structures, and research that help protect our core values.

II. THE OBAMA ADMINISTRATION'S APPROACH TO OPEN DATA AND PRIVACY

Throughout American history, technology and privacy laws have evolved in tandem. The United States has long been a leader in protecting individual privacy while supporting an environment of innovation and economic prosperity.

The Fourth Amendment to the Constitution protects the "right of the people to be secure in their persons, houses, papers, and effects, against unreasonable searches and seizures." Flowing from this protection of physical spaces and tangible assets is a broader sense of respect for security and dignity that is indispensable both to personal well-being and to the functioning of democratic society.[23] A legal framework for the protection of privacy interests has grown up in the United States that includes constitutional, federal, state, and common law elements. "Privacy" is thus not a narrow concept, but instead addresses a range of concerns reflecting different types of intrusion into a person's sense of self, each requiring different protections.

Data collection—and the use of data to serve the public good—has an equally long history in the United States. Article I, Section 2 of the Constitution mandates a decennial Census in order to apportion the House of Representatives. In practice, the Census has never been conducted as just a simple head count, but has always been used to determine more specific demographic information for public purposes.[24]

Since President Obama took office, the federal government has taken unprecedented steps to make more of its own data available to citizens, companies, and innovators. Since 2009, the Obama Administration has made tens of thousands of datasets public, hosting many of them on Data.gov, the central clearinghouse for U.S. government data. Treating government data as an asset and making it available, discoverable, and usable—in a word, open—strengthens democracy, drives economic opportunity, and improves citizens' quality of life.

Deriving value from open data requires developing the tools to understand and analyze it. So the Obama Administration has also made significant investments in the basic science of data analytics, storage, encryption, cybersecurity, and computing power.

The Obama Administration has made these investments while also recognizing that the collection, use, and sharing of data pose serious challenges. Federal research dollars have supported work to address the technological and ethical issues that arise when handling large-scale data sets. Drawing on the United States' long history of leadership on privacy issues, the Obama Administration also issued a groundbreaking consumer privacy blueprint in 2012 that included a Consumer Privacy Bill of Rights.[25] In 2014, the President announced the Cybersecurity Framework, developed in partnership with the private sector, to strengthen the security of the nation's critical infrastructure.[26]

This section charts the intersections of these initiatives—ongoing efforts to harness data for the public good while ensuring the rights of citizens and consumers are protected.

Open Data in the Obama Administration

Open Data Initiatives

The smartphones we carry around in our pockets tell us where we are by drawing on open government data. Decades ago, the federal government first made meteorological data and the Global Positioning System freely available, enabling entrepreneurs to create a wide range of new tools and services, from weather apps to automobile navigation systems.

In the past, data collected by the government mostly stayed in the government agency that collected it. The Obama Administration has launched a series of Open Data Initiatives, each unleashing troves of valuable data that were previously hard to access, in domains including health, energy, climate, education, public safety, finance, and global development. Executive Order 13642, signed by President Obama on May 9, 2013, established an important new principle in federal stewardship of data: going forward, agencies must consider openness and machine-readability as the new defaults for government information, while appropriately safeguarding privacy, confidentiality, and security.[27] Extending these open data efforts is also a core element of the President's Second Term Management Agenda, and the Office of Management and Budget has directed agencies to release more of the administrative information they use to make decisions so it might be useful to others.[28]

At Data.gov the public can find everything from data regarding complaints made to the federal Consumer Financial Protection Bureau about private student loans to 911 service area boundaries for the state of Arkansas. The idea is that anyone can use Data.gov to find the open data they are looking for without having specialized knowledge of government agencies or programs within those agencies. Interested software developers can use simple tools to automatically access the datasets.

Federal agencies must also prioritize their data release efforts in part based on requests from the public. Each agency is required to solicit input through digital feedback mechanisms, like an email address or an online platform. For the first time, any advocate, entrepreneur, or researcher can connect with the federal government and suggest what data should be made available. To further improve feedback and encourage productive use of open government data, Administration officials have hosted and participated in a range of code-a-thons, brainstorming workshops ("Data Jams"), showcase events ("Datapaloozas"), and other meetings about open government data.[29]

Pursuant to the May 2013 Executive Order, the Office of Management and Budget and the Office of Science and Technology Policy released a framework for agencies to manage information as an asset throughout its lifecycle, which includes requirements to continue to protect personal, sensitive, and confidential data.[30] Agencies already categorize data assets into three access levels—public, restricted public, and non-public—and publish only the public catalog. To promote transparency, agencies include information in their external data inventories about technically public data assets that have not yet been posted online.

My Data Initiatives

Making public government data more open and machine-readable is only one element of the Administration's approach to data. The Privacy Act of 1974 grants citizens certain rights of access to their personal information. That access should be easy, secure, and useful. Starting in 2010, the Obama Administration launched a series of My Data initiatives to empower Americans with secure access to their personal data and increase citizens' access to private-sector applications and services that can be used to analyze it. The My Data initiatives include:

- **Blue Button:** The Blue Button allows consumers to securely access their health information so they can better manage their health care and finances and share their information with providers. In 2010, the U.S. Department of Veterans Affairs launched the Blue Button to give veterans the ability to download their health records. Since then, more than 5.4 million veterans have used the Blue Button tool to access their personal health information. More than 500 companies in the private sector have pledged their support to increase patient access to

their health data by leveraging Blue Button, and today, more than 150 million Americans have the promise of being able to access their digital health information from health care providers, medical laboratories, retail pharmacy chains, and state immunization registries.
- **Get Transcript:** In 2014, the Internal Revenue Service made it possible for taxpayers to digitally access their last three years of tax information through a tool called Get Transcript. Individual taxpayers can use Get Transcript to download a record of past tax returns, which makes it easier to apply for mortgages, student loans, and business loans, or to prepare future tax filings.
- **Green Button:** The Administration partnered with electric utilities in 2012 to create the Green Button, which provides families and business with easy access to their energy usage information in a consumer-friendly and computer-friendly format. Today, 48 utilities and electricity suppliers serving more than 59 million homes and businesses have committed to giving their customers "Green Button" access to help them save energy. With customers in control of their energy data, they can choose which private sector tools and services can help them better manage their property's energy efficiency.[31]
- **MyStudentData:** The Department of Education makes it possible for students and borrowers to access and download their data from the Free Application for Federal Student Aid and their federal student loan information—including loan, grant, enrollment, and overpayment information. In both cases, the information is available via a user-friendly, machine-readable, plain-text file.

Beyond providing people with easy and secure access to their data, the My Data initiatives helps establish a strong model for personal data accessibility that the Administration hopes will become widely adopted in the private and public sectors. The ability to access one's personal information will be increasingly important in the future, when more aspects of life will involve data transactions between individuals, companies, and institutions.

Big Data Initiative: "Data to Knowledge to Action"

At its core, big data is about being able to move quickly from data to knowledge to action. On March 29, 2012, six federal agencies joined forces to launch the "Big Data Research and Development Initiative," with over $200 million in research funding to improve the tools and techniques needed to access, organize, and glean discoveries from huge volumes of digital data.

Since the launch of this "Data to Knowledge to Action" initiative, DARPA has created an "Open Catalog" of the research publications and open source software generated by its $100 million XDATA program, an effort to process and analyze large sets of imperfect, incomplete data.[32] The National Institutes of Health has supported a $50 million "Big Data to Knowledge" program about biomedical big data. The National Science Foundation has funded big data research projects which have reduced the cost of processing a human genome by a factor of 40. The Department of Energy announced a $25 million Scalable Data Management, Analysis, and Visualization Institute, which produced climate data techniques that have made seasonal hurricane predictions more than 25 percent more accurate. Many other research initiatives have important big data components, including the BRAIN Initiative, announced by President Obama in April 2013. As part of the Administration's big data research initiative, the National Science Foundation has also funded specific projects examining the social, ethical, and policy aspects of big data.

U.S. Privacy Law and International Privacy Frameworks

Development of Privacy Law in the United States

U.S. privacy laws have shaped and been shaped by societal changes, including the waves of technological innovation set in motion by the industrial revolution. The first portable cameras helped catalyze Samuel Warren and Louis Brandeis's seminal 1890 article *The Right to Privacy*, in which they note that "[r]ecent inventions and business methods call attention to the next step which must be taken for the protection of the person, and for securing to the individual ... the right 'to be let alone'... numerous mechanical devices threaten to make good the prediction that 'what is whispered in the closet shall be proclaimed from the house-tops.'"[33] This prescient work laid the foundation for the common law of privacy in the 20th century, establishing citizens' rights to privacy from the government and from each other.[34]

Over the course of the last century, case law about what constitutes a "search" for purposes of the Fourth Amendment to the Constitution has developed with time and technology.[35] In 1928, the U.S. Supreme Court held in *Olmstead v. United States* that placing wiretaps on a phone line located outside of a person's house did not violate the Fourth Amendment, even

Big Data: Seizing Opportunities, Preserving Values 17

though the government obtained the content from discussions *inside* the home.[36] But the *Olmstead* decision was arguably more famous for the dissent written by Justice Brandeis, who wrote that the Founders had "conferred, as against the government, the right to be let alone—the most comprehensive of rights and the right most favored by civilized men."[37]

The Court's opinion in *Olmstead* remained the law of the land until it was overturned by the Court's 1967 decision in *Katz v. United States*. In *Katz*, the Court held that the FBI's placement of a recording device on the outside of a public telephone booth without a warrant qualified as a search that violated the "reasonable expectation of privacy" of the person using the booth, even though the device did not physically penetrate the booth, his person, or his property. Under *Katz*, an individual's subjective expectations of privacy are protected when society regards them as reasonable.[38]

Civil courts did not immediately acknowledge privacy as justification for one citizen to bring a lawsuit against another—what lawyers call a "cause of action." It wasn't until the 1934 Restatement (First) of Torts that an "unreasonable and serious" invasion of privacy was recognized as a basis to sue.[39] Courts in most states began to recognize privacy as a cause of action, although what emerged from decisions was not a single tort, but instead "a complex of four" potential torts:[40]

1. Intrusion upon a person's seclusion or solitude, or into his private affairs.
2. Public disclosure of embarrassing private facts about an individual.
3. Publicity placing one in a false light in the public eye.
4. Appropriation of one's likeness for the advantage of another.[41]

Some contemporary critics argue the "complex of four" does not sufficiently recognize privacy issues that arise from the extensive collection, use, and disclosure of personal information by businesses in the modern marketplace. Others suggest that automated processing should in fact ease privacy concerns because it uses computers operated under precise controls to perform tasks that used to be handled by a person.[42]

The Fair Information Practice Principles

As computing advanced and became more widely used by government and the private sector, policymakers around the world began to tackle the issue of privacy anew. In 1973, the U.S. Department of Health, Education, and Welfare issued a report entitled *Records, Computers, and the Rights of*

Citizens.[43] The report analyzed "harmful consequences that might result from automated personal data systems" and recommended certain safeguards for the use of information. Those safeguards, commonly known today as the "Fair Information Practice Principles," or "FIPPs," form the bedrock of modern data protection regimes.

While the principles are instantiated in law and international agreements in different ways, at their core, the FIPPs articulate basic protections for handling personal data. They provide that an individual has a right to know what data is collected about him or her and how it is used. The individual should further have a right to object to some uses and to correct inaccurate information. The organization that collects information has an obligation to ensure that the data is reliable and kept secure. These principles, in turn, served as the basis for the Privacy Act of 1974, which regulates the federal government's maintenance, collection, use, and dissemination of personal information in systems of records.[44]

By the late 1970s, several other countries had also passed national privacy laws.[45] In 1980, the Organization for Economic Cooperation and Development (OECD) issued its "Guidelines Governing the Protection of Privacy and Transborder Flow of Personal Data."[46] Building on the FIPPs, the OECD guidelines have informed national privacy laws, sector-specific laws, and best practices for the past three decades. In 1981, the Council of Europe also completed work on the Convention for the Protection of Individuals with regard to Automatic Processing of Personal Data (Convention 108), which applied a FIPPs approach to emerging privacy concerns in Europe.

Despite some important differences, the privacy frameworks in the United States and those countries following the EU model are both based on the FIPPs. The European approach, which is based on a view that privacy is a fundamental human right, generally involves top-down regulation and the imposition of across-the-board rules restricting the use of data or requiring explicit consent for that use. The United States, in contrast, employs a sectoral approach that focuses on regulating specific risks of privacy harm in particular contexts, such as health care and credit. This places fewer broad rules on the use of data, allowing industry to be more innovative in its products and services, while also sometimes leaving unregulated potential uses of information that fall between sectors.

The FIPPs form a common thread through these sectoral laws and a variety of international agreements. They are woven into the 2004 Asia Pacific Economic Cooperation Privacy Principles, which was endorsed by APEC economies, and form the basis for the U.S.-E.U. and U.S.-Switzerland Safe

Harbor Frameworks, which harness the global consensus around the FIPPs as a means to build bridges between U.S. and European law.[47]

Sector-Specific Privacy Laws in the United States

In the United States during the 1970s and 80s, narrowly-tailored sectoral privacy laws began to supplement the tort-based body of common law. These sector-specific laws create privacy safeguards that apply only to specific types of entities and data. With a few exceptions, individual states and the federal government have predominantly enacted privacy laws on a sectoral basis.[48]

The Fair Credit Reporting Act (FCRA) was originally enacted in 1970 to promote accuracy, fairness, and privacy protection with regard to the information assembled by consumer reporting agencies for use in credit and insurance reports, employee background checks, and tenant screenings. The law protects consumers by providing specific rights to access and correct their information. It requires companies that prepare consumer reports to ensure data is accurate and complete; limits when such reports may be used; and requires agencies to provide notice when an adverse action, such as the denial of credit, is taken based on the content of a report.

The 1996 Health Insurance Portability and Accountability Act (HIPAA) addresses the use and disclosure of individuals' health information by specified "covered entities" and includes standards designed to help individuals understand and control how their health information is used.[49] A key aspect of HIPAA is the principle of "minimum necessary" use and disclosure.[50] Congress and the Department of Health and Human Services have periodically updated protections for personal health data. The Children's Online Privacy Protection Act of 1998 (COPPA) and the Federal Trade Commission's implementing regulations require online services directed at children under the age of 13, or which collect personal data from children, to obtain verifiable parental consent to do so. In the financial sector, the Gramm-Leach-Bliley Act mandates that financial institutions respect the privacy of customers and the security and confidentiality of those customers' nonpublic personal information. Other sectoral privacy laws safeguard individuals' educational, communications, video rental, and genetic information.[51]

Consumer Privacy Bill of Rights

In February 2012, the White House released a report titled *Consumer Data Privacy in a Networked World: A Framework for Protecting Privacy and Promoting Innovation in the Global Digital Economy*.[52] This "Privacy Blueprint" contains four key elements: a Consumer Privacy Bill of Rights based on the Fair Information Practice Principles; a call for government-convened multi-stakeholder processes to apply those principles in particular business contexts; support for effective enforcement of privacy rights, including the enactment of baseline consumer privacy legislation; and a commitment to international privacy regimes that support the flow of data across borders.

At the center of the Privacy Blueprint is the Consumer Privacy Bill of Rights, which states clear baseline protections for consumers. The rights are:

- **Individual Control:** Consumers have a right to exercise control over what personal data organizations collect from them and how they use it.
- **Transparency:** Consumers have a right to easily understandable information about privacy and security practices.
- **Respect for Context:** Consumers have a right to expect that organizations will collect, use, and disclose personal data in ways that are consistent with the context in which consumers provide the data.
- **Security:** Consumers have a right to secure and responsible handling of personal data.
- **Access and Accuracy:** Consumers have a right to access and correct personal data in usable formats, in a manner that is appropriate to the sensitivity of the data and the risk of adverse consequences to consumers if the data are inaccurate.
- **Focused Collection:** Consumers have a right to reasonable limits on the personal data that companies collect and retain.
- **Accountability:** Consumers have a right to have personal data handled by companies with appropriate measures in place to assure they adhere to the Consumer Privacy Bill of Rights.

The Consumer Privacy Bill of Rights is more focused on consumers than previous privacy frameworks, which were often couched in legal jargon. For example, it describes a right to "access and accuracy," which is more easily understood by users than previous formulations referencing "data quality and integrity." Similarly, it assures consumers that companies will respect the "context" in which data is collected and used, replacing the term "purpose specification."

The Consumer Privacy Bill of Rights also draws upon the Fair Information Practice Principles to better accommodate the online environment in which we all now live. Instead of requiring companies to adhere to a single, rigid set of requirements, the Consumer Privacy Bill of Rights establishes general principles that afford companies discretion in how they implement them. The Consumer Privacy Bill of Rights' "context" principle interacts with its other six principles, assuring consumers that their data will be collected and used in ways consistent with their expectations. At the same time, the context principle permits companies to develop new services using personal information when that use is consistent with the companies' relationship with its users and the circumstances surrounding how it collects data.

The Internet's complexity, global reach, and constant evolution require timely, scalable, and innovation-enabling policies. To answer this challenge, the Privacy Blueprint calls for all relevant stakeholders to come together to develop voluntary, enforceable codes of conduct that specify how the Consumer Privacy Bill of Rights applies in specific business contexts. The theory behind the Consumer Privacy Bill of Rights is that this combination of broad baseline principles and specific codes of conduct can protect consumers while supporting innovation.

Promoting Global Interoperability

The Obama Administration released the Consumer Privacy Bill of Rights as other countries and international organizations began to review their own privacy frameworks. In 2013, the OECD updated its Privacy Guidelines, which supplement the Fair Information Practice Principles with mechanisms to implement and enforce privacy protections. The APEC Cross Border Privacy Rules System, also announced in 2013, largely follows the OECD guidelines.[53] The Council of Europe is undertaking a review of Convention 108. Building bridges among these different privacy frameworks is critical to ensuring robust international commerce.

The European Union is also in the process of reforming its data protection rules.[54] The current E.U. Data Protection Directive only allows transfers of E.U. citizens' data to those non-E.U. countries with "adequate" privacy laws or mechanisms providing sufficient safeguards for data, such as the U.S.-E.U. Safe Harbor. In January 2014, the U.S. and E.U. began discussing how best to enhance the Safe Harbor Framework to ensure that it continues to provide strong data protection and enable trade through increased transparency, effective enforcement, and legal certainty. These negotiations continue, even as Europe—like the United States—wrestles with questions about how it will accommodate big data technologies and increased computational and storage capacities.[55]

In March 2014, the Federal Trade Commission, together with agency officials from the European Union and Asia-Pacific Economic Cooperation economies, announced joint E.U. and APEC endorsement of a document that maps the requirements of the European and APEC privacy frameworks.[56] The mapping project will help companies seeking certification to do business in both E.U. and APEC countries recognize overlaps and gaps between the two frameworks.[57] Efforts like these clarify obligations for companies and help build interoperability between global privacy frameworks.

Conclusion

The most common privacy risks today still involve "small data"—the targeted compromise of, for instance, personal banking information for purposes of financial fraud. These risks do not involve especially large volumes, rapid velocities, or great varieties of information, nor do they implicate the kind of sophisticated analytics associated with big data. Protecting privacy of "small" data has been effectively addressed in the United States through the Fair Information Practice Principles, sector-specific laws, robust enforcement, and global privacy assurance mechanisms.

Privacy scholars, policymakers, and technologists are now turning to the question of how big data technology can be effectively managed under the FIPPs-based frameworks. The remainder of this report explores applications of big data in the public and private sector and then returns to consider the overall implications big data may have on current privacy frameworks.

III. PUBLIC SECTOR MANAGEMENT OF DATA

Government keeps the peace. It makes sure our food is safe to eat. It keeps our air and water clean. The laws and regulations it promulgates order economic and political life. Big data technology stands to improve nearly all the services the public sector delivers.

This section explores how big data is already helping the government carry out its obligations in health, education, homeland security, and law enforcement. It also begins to frame some of the challenges big data raises. Questions about what the government should and should not do, and how the rights of citizens should be protected in light of changing technology, are as old as the Republic itself. In framing the laws and norms of our young country, the founders took pains to demarcate private spheres shielded from inappropriate government interference. While many things about the big data world might astonish them, the founders would not be surprised to find that the Constitution and Bill of Rights are as central to the debate as Moore's law and zettabytes.

At its core, public-sector use of big data heightens concerns about the balance of power between government and the individual. Once information about citizens is compiled for a defined purpose, the temptation to use it for other purposes can be considerable, especially in times of national emergency. One of the most shameful instances of the government misusing its own data dates to the Second World War. Census data collected under strict guarantees of confidentiality was used to identify neighborhoods where Japanese-Americans lived so they could be detained in internment camps for the duration of the war.

Because the government bears a special responsibility to protect its citizens when exercising power and authority for the public good, how big data should be put to use in the public sector, as well as what controls and limitations should apply, must be carefully considered. If unchecked, big data could be a tool that substantially expands government power over citizens. At the same time, big data can also be used to enhance accountability and to engineer systems that are inherently more respectful of privacy and civil rights.

Big Data and Health Care Delivery

Data has long been a part of health care delivery. In the past several years, legislation has created incentives for health care providers to transition to using electronic health records, vastly expanding the volume of health data available to clinicians, researchers, and patients. With the enactment of the Affordable Care Act, the model for health care reimbursement is beginning to shift from paying for isolated and potentially uncoordinated instances of treatment—a model called "fee-for-service"—to paying on the basis of better health outcomes. Taken together, these trends are helping build a "learning" health care system where effective practices are identified from clinical data and then rapidly disseminated back to providers.

Big data can identify diet, exercise, preventive care, and other lifestyle factors that help keep people from having to seek care from a doctor. Big data analytics can also help identify clinical treatments, prescription drugs, and public health interventions that may not appear to be effective in smaller samples, across broad populations, or using traditional research methods. From a payment perspective, big data can be used to ensure professionals who treat patients have strong performance records and are reimbursed on the quality of patient outcomes rather than the quantity of care delivered.

The emerging practice of predictive medicine is the ultimate application of big data in health. This powerful technology peers deeply into a person's health status and genetic information, allowing doctors to better predict whether individuals will develop a disease and how they might respond to specific therapies. Predictive medicine raises many complex issues. Traditionally, health data privacy policies have sought to protect the identity of individuals whose information is being shared and analyzed. But increasingly, data about groups or categories of people will be used to identify diseases prior to or very early after the onset of clinical symptoms.

But the information that stands to be discovered by predictive medicine extends beyond a single individual's risks to include others with similar genes, potentially including the children and future descendants of those whose information is originally collected. Bio-repositories that link genomic data to health care data are on the leading edge of confronting important questions about personal privacy in the context of health research and treatment.[58]

The privacy frameworks that currently cover information now used in health may not be well suited to address these developments or facilitate the research that drives them. Using big data to improve health requires advanced analytical models to ingest multiple kinds of lifestyle, genomic, medical, and financial data. The powerful connection between lifestyle and health outcomes means the distinction between personal data and health care data has begun to blur. These types of data are subjected to different and sometimes conflicting federal and state regulation, including the Health Insurance Portability and Accountability Act, Gramm-Leach-Bliley Act, Fair Credit Reporting Act, and Federal Trade Commission Act. The complexity of complying with numerous laws when data is combined from various sources raises the potential need to carve out special data use authorities for the health care industry if it is to realize the potential health gains and cost reductions that could come from big data analytics. At the same time, health organizations interact with many organizations that are not regulated under any of these laws.[59] In the resulting ecosystem, personal health information of various kinds is shared with an array of firms, and even sold by state governments, in ways that might not accord with consumer expectations of the privacy of their medical data.

Though medicine is changing, information about our health remains a very private part of our lives. As big data enables ever more powerful discoveries, it will be important to revisit how privacy is protected as information circulates among all the partners involved in care. Health care leaders have voiced the need for a broader trust framework to grant all health information, regardless of its source, some level of privacy protection. This may potentially involve crafting additional protections beyond those afforded in the Health Insurance Portability and Accountability Act and Genetic Information Non-Discrimination Act as well as streamlining data interoperability and compliance requirements. After studying health information technology, the President's Council of Advisors on Science & Technology concluded that the nation needs to adopt universal standards and an architecture that will facilitate controlled access to information across many different types of records.[60]

Modernizing the health care data privacy framework will require careful negotiation between the many parties involved in delivering health care and insurance to Americans, but the potential economic and health benefits make it well worth the effort.

Learning about Learning: Big Data and Education

Education at both the K-12 and university levels is now supported inside and outside the classroom by a range of technologies that help foster and enhance the learning process. Students now access class materials, watch instructional videos, comment on class activities, collaborate with each other, complete homework, and take tests online.

Technology-based educational tools and platforms offer important new capabilities for students and teachers. After only a few generations of evolution, these tools provide real-time assessment so that material can be presented based on how quickly a student learns. Education technologies can also be scaled to reach broad audiences, enable continuous improvement of course content, and increase engagement among students.[61]

Beyond personalizing education, the availability of new types of data profoundly improves researchers' ability to learn about learning. Data from a student's experience in massive open online courses (MOOCs) or other technology-based learning platforms can be precisely tracked, opening the door to understanding how students move through a learning trajectory with greater fidelity, and at greater scale, than traditional education research is able to achieve. This includes gaining insight into student access of learning activities, measuring optimal practice periods for meeting different learning objectives, creating pathways through material for different learning approaches, and using that information to help students who are struggling in similar ways. Already, the Department of Education has studied how to harness these technologies, begun integrating the use of data from online education in the National Education Technology Plan, and laid plans for a Virtual Learning Lab to pioneer the methodological tools for this research.[62]

The big data revolution in education also raises serious questions about how best to protect student privacy as technology reaches further into the classroom. While states and local communities have traditionally played the dominant role in providing education, much of the software that supports online learning tools and courses is provided by forprofit firms. This raises complicated questions about who owns the data streams coming off online education platforms and how they can be used. Applying privacy safeguards like the Family Educational Rights and Privacy Act, the Protection of Pupil Rights Amendment, or the Children's Online Privacy Protection Act to educational records can create unique challenges.

PROTECTING CHILDREN'S PRIVACY IN THE ERA OF BIG DATA

Children today are among the first generation to grow up playing with digital devices even before they learn to read. In the United States, children and teenagers are active users of mobile apps and social media platforms. As they use these technologies, granular data about them—some of it sensitive—is stored and processed online. This data has the potential to dramatically improve learning outcomes and open new opportunities for children, but could be used to build an invasive consumer profile of them once they become adults, or otherwise pose problems later in their lives. Although youth on average are typically no less, and in many cases more, cognizant of commercial and government use of data than adults, they often face scrutiny by parents, teachers, college admissions officers, military recruiters, and case workers. Vulnerable youth, including foster children and homeless youth, who typically have little adult guidance, are also particularly susceptible to data misuse and identity theft. Struggling to find some privacy in the face of tremendous supervision, many youth experiment with various ways to obscure the meaning of what they share except to select others, even if they are unable to limit access to the content itself.[63]

> Because young people are exactly that—young—they need appropriate freedoms to explore and experiment safely and without the specter of being haunted by mistakes in the future. The Children's Online Privacy Protection Act requires website operators and app developers to gain consent from a parent or guardian before collecting personal information from children under the age of 13. There is not yet a settled understanding of what harms, if any, are accruing to children and what additional policy frameworks may be needed to ensure that growing up with technology will be an asset rather than a liability.

Just as with health care, some of the information revealed when a user interacts with a digital education platform can be very personal, including aptitude for particular types of learning and performance relative to other students. It is even possible to discern whether students have learning disabilities or have trouble concentrating for long periods. What time of day and for how long students stay signed in to online tools reveals lifestyle habits. What should educational institutions do with this data to improve learning opportunities for students? How can students who use these platforms, especially those in K-12 education, be confident that their data is safe?

To help answer complicated questions about ownership and proper usage of data, the U.S. Department of Education released guidance for online education services in February 2014.[64] This guidance makes clear that schools and districts can enter into agreements with third parties involving student data only so long as requirements under the Family Educational Rights and Privacy Act and Protection of Pupil Rights Amendment are met. As more online learning tools and services become available for kids, states and local governments are also watching these issues closely.[65] Schools and districts can only share protected student information to further legitimate educational interests, and they must retain "direct control" over that information. Even with this new guidance, the question of how best to protect student privacy in a big data world must be an ongoing conversation.

The Administration is committed to vigorously pursuing these questions and will work through the Department of Education so all students can experience the benefits of big data innovations in teaching and learning while being protected from potential harms.[66] As Secretary of Education Arne Duncan has said, "Student data must be secure, and treated as precious, no matter where it's stored. It is not a commodity."[67] This means ensuring the personal information and online activity of students are protected from inappropriate uses, especially when it is gathered in an educational context.

Big Data at the Department of Homeland Security

Every day, two million passengers fly into, within, or over the United States. More than a million people enter the country by land. Verifying the identity of each person and determining whether he or she poses a threat falls to the Department of Homeland Security, which must process huge amounts of data in seconds to carry out its mission. The Department is not simply out to find the "needle in the haystack." Protecting the homeland often depends on finding the most critical needles across many haystacks—a classic big data problem.

Ensuring the Department efficiently and lawfully uses the information it collects is a massive undertaking. DHS was created out of 22 separate government agencies in the wake of the 9/11 attacks. Many of the databases DHS operates today are physically disconnected, run legacy operating systems, and are unable to integrate information across different security classifications. The Department also carries out a diverse portfolio of missions, each governed by separate authorities in law. At all times, information must be used only for authorized purposes and in ways that protect the privacy and civil liberties afforded to U.S. citizens and foreign nationals who enter or reside in the United States. Ensuring information is properly used falls to six offices at DHS headquarters.

Beginning in 2012, representatives of the Chief Information Officer, the policy division, and the intelligence division came together with privacy, civil liberties and legal oversight officers to begin developing the first department-wide big data capability, resident in two pilot programs named Neptune and Cerberus.[68] Neptune is designed from the ground up to be a "data lake" into which unclassified information from different sources flows.[69] It has multiple built-in safeguards, including the ability to apply multiple data tags and finegrained rules to determine which users can access which data for what purpose. All of the data is tagged according to a precise scheme. The rules governing usage focus on whether there is an authorized purpose, mission, or "need to know," and whether the user has the appropriate job series and clearance to access the information. In this way, data tags can be combined with user attributes and context to govern what information is used where and by whom.

The Neptune and Cerberus pilots also contain important controls around the types of searches that users are permitted to perform. A primary inspection agent may only need to perform a search on a specific person, because the agent is trying to confirm basic biographical information. However, an

Immigration and Customs Investigator may need to perform person and characteristic searches while investigating a crime. DHS intelligence analysts may need to perform searches based on identities, characteristics, and trends when analyzing information related to a threat to homeland security. System administrators have no need to access the data contained within the system. The architecture of the database allows them to maintain the overall IT system but not to access any individual records.

A MODEL FOR MANAGING DATA

To build the tagging standards that govern information in its big data pilots, the Department of Homeland Security brought together the owners of the data systems, called data stewards, with representatives from privacy, civil liberties, and legal oversight offices. For each database field, the group charted its attributes and how access to the data is granted to different user communities. After developing a set of tags to encode this information, they then considered what additional rules and protections were needed to account for specific use limitations or special cases governed by law or regulation. Tagging both enables precise access control and preserves links to source data and the purpose of its original collection. The end result is a taxonomy of rules governing where information goes and tracking where it came from and under what authority.

The fields in each database are grouped into three categories: core biographical data, such as name, date of birth, and citizenship status; extended biographical data, including addresses, phone number, and email; and detailed encounter data derived from electronic and in-person interactions with DHS. Encounter data is the most sensitive category. It may contain a law enforcement officer's observations about an individual they interview as well as allegations of a risk to homeland security they may pose. These data tags then allow precise rules to be set of who can access what information for what reason. In these two pilots, the majority of rules for negotiating access are consistent across DHS's different user communities. For example, many users will need access to the core biographic information of a particular data set to perform their missions. But some of the rules require far greater customization to account for specific use limitations.

The capabilities developed in these pilots are of a whole different order than the databases DHS inherited in 2002. Before these big data initiatives, it was not easy to perform searches across databases held by different components, let alone to aggregate them. In the past, users and system administrators might have been issued a login and username and granted total access, sometimes without an audit trail monitoring their use. Now, DHS will be able to more precisely grant access according to mission needs. Most importantly, by being deliberate in tagging and organizing the data in these advanced repositories, the agency can take on new kinds of predictive and anomaly analysis while complying with the law and subjecting its activities to robust oversight.

It's no accident that DHS was able to so carefully engineer how data is handled. DHS has both a dedicated Privacy Office and an Office for Civil Rights and Civil Liberties, each staffed with experts to help navigate this complex terrain.[70] Each pilot is accompanied by a detailed privacy impact assessment released to the public in advance of its operation. DHS has provided public briefings on the pilots and allowed members of the public to ask questions about the initiatives. The privacy and civil liberties oversight officials not only approved the plan for the pilots, they also approve tools or widgets built in the future to increase their functionality. All of this helps drive improvements to DHS's mission while ensuring that privacy and civil liberties concerns are considered from the start.

Upholding Our Privacy Values in Law Enforcement

Big data can be a powerful tool for law enforcement. Recently, advanced web tools developed by DARPA's Memex program have helped federal law enforcement make substantial progress in identifying human trafficking networks in the United States. These tools comb the "surface web" we all know, as well as "deep web" pages that are also public but not indexed by commonly used search engines. By allowing searches across a wide range of websites, the tools uncover a wealth of information that might otherwise be difficult or time-intensive to obtain. Possible trafficking rings can be identified and cross-referenced with existing law enforcement databases, helping police officers map connections between sex trafficking and other illegal activity. Already, the tools have helped detect trafficking networks originating in Asia and spreading to several U.S. cities. It's a powerful example of how big data can help protect some of the most vulnerable people in the world.

Big data technologies provide effective tools to law enforcement and other agencies that protect our security, but they also pose difficult questions about their appropriate uses. Blending multiple data sources can create a fuller picture of a suspect's activities around the time of a crime, but can also aid in the creation of suspect profiles that focus scrutiny on particular individuals with little or no human intervention. Pattern analysis can reveal how criminal organizations are structured or can be used to make predictions about possible future crimes. Gathering broad datasets can help catch criminals, but can also sweep up detailed personal information about people who are not subjects of an investigation. When it comes to law enforcement, we must be careful to ensure that big data technologies are used in ways that take into account the needs to protect public safety and fairly enforce the laws, as well as the civil liberties and legitimate privacy interests of citizens.

Big data will naturally—and appropriately—be used differently in national security. A powerful intelligence system that harnesses global data to identify terrorist networks, to provide warning of impending attacks, and to prevent the proliferation of weapons of mass destruction will operate under different legal authorities and oversight and have different privacy protections than a law enforcement system that helps allocate police resources to neighborhoods where higher levels of crime are predicted. Even though the applications are different, there are nevertheless important similarities in how privacy and civil rights are maintained across law enforcement and intelligence contexts. Privacy and legal officials must certify use of a system in each case, minimization rules are often employed to reduce information held, and data-tagging techniques are used to control access.

New Tools and New Challenges

The use of new technologies, especially in law enforcement, has given rise to important Constitutional jurisprudence.[71] As Justice Alito observed in a 2013 Supreme Court case concerning police placement of a GPS tracker on a suspect's car without a court order: "[I]t is almost impossible to think of late-18th-century situations that are analogous to what took place in this case. (Is it possible to imagine a case in which a constable secreted himself somewhere in a coach and remained there for a period of time in order to monitor the movements of the coach's owner?"[72] Alito noted further, "Something like this might have occurred in 1791, but this would have required either a gigantic coach, a very tiny constable, or both.)"[73]

The "tiny constable" has enormous implications. Ubiquitous surveillance—whether by GPS tracking, closed circuit TV, or virtually undetectable sensors—will increasingly figure in litigation about reasonable expectations of privacy and the proper uses and limits of law enforcement technology.

In recent decades, the cost of surveillance and the physical size of surveillance equipment have rapidly decreased. This has made it feasible for over 70 cities in the United States to install audio sensors that can pinpoint gunfire and rapidly dispatch police to a potential crime scene.[74] Given the speed of access and decreasing cost of storage, it has likewise become practical for even local police forces to actively collect and catalog data, like license plate and vehicle information, in real-time on a city-wide scale, and to also retain it for later use.[75]

The benefits of some of these technologies are tremendous. From finding missing persons to launching complex manhunts, the use of advanced surveillance technology by federal, state, and local law enforcement can mean a faster and more effective response to criminal activity. It can also increase the chances that justice is reliably served in online crime, where criminals are among the earliest adopters of new technologies and law enforcement needs to have timely access to digital evidence.

Beyond surveillance, predictive technologies offer the potential for law enforcement to be better prepared to anticipate, intervene in, or outright prevent certain crimes. Some analytics software, such as one program in use by both the Los Angeles and Memphis police departments, employs predictive analytics to identify geographically-based "hotspots."[76] Many cities attribute meaningful declines in property crime to stepping up police patrols in "hotspot" areas.

Controversially, predictive analytics can now be applied to analyze a person's individual propensity to criminal activity.[77] In response to an epidemic of gang-related murders, the city of Chicago conducted a pilot that shifts the focus of predictive policing from geographical factors to identity. By drawing on police and other data and applying social network analysis, the Chicago police department assembled a list of roughly 400 individuals identified by certain factors as likely to be involved in violent crime. As a result, police have a heightened awareness of particular individuals that might reflect factors beyond charges and convictions that are part of the public record.[78]

Predictive analytics are also being used in other areas of criminal justice. In Philadelphia, police are using software designed to predict which parolees are more likely to commit a crime after release from prison and thus should have greater supervision.[79] The software uses about two dozen variables, including age, criminal history, and geographic location.

These new techniques have come with considerable controversy about how and when they should be deployed.[80] This technology can help more precisely allocate law enforcement and other public resources, which can lead to the prevention of harmful crimes. At the same time, our Constitution and Bill of Rights grant certain rights that must not be abridged.

Police departments' potential use of a new array of data and algorithms to try to predict criminal propensities and redirect police powers in advance of criminal activity has important consequences. It requires careful review of how we define "individualized suspicion," which is the constitutional predicate of surveillance and search.[81] The presence and persistence of authority, and the reasonable belief that one's activities, movements, and personal affiliations are being monitored by law enforcement, can have a chilling effect on rights of free speech and association. The next section considers where changes in technology introduce tension within particular areas of the law.

Implications of Big Data Technology for Privacy Law

Access to Data Held by Third Parties

Personal documents and records have evolved from paper kept in the home, to electronic files held on the hard drive of a computer in the home, to many different kinds of computer files kept both locally and in cloud repositories accessed across multiple devices within and outside the home. As remote processing and cloud storage technologies increasingly become the norm for personal computing and records management, we must take measure of the how the law accounts for these developments.

Whether an individual reasonably expects an act to be private has framed much of our thinking about what protections are deserved. As Justice Potter Stewart in the 1967 Katz majority opinion noted: "[T]he Fourth Amendment protects people, not places. What a person knowingly exposes to the public, even in his own home or office, is not a subject of Fourth Amendment protection...But what he seeks to preserve as private, even in an area accessible to the public, may be constitutionally protected."[82]

Two later Supreme Court decisions further elaborated on how the Fourth Amendment applies to information that is shared with third parties. In *United States v. Miller*, in 1976, the Court found that the Fourth Amendment does not prohibit the government from obtaining "information revealed to a third-party and conveyed by him to government authorities, even if the information is revealed on the assumption that it will be used only for a limited purpose and the confidence placed in the third-party will not be betrayed."[83] Three years later, the Supreme Court held in *Smith v. Maryland* that the telephone numbers a person dials are not protected by a reasonable expectation of privacy because the caller voluntarily conveys dialing information to the phone company. The Court again affirmed that it had "consistently . . . held that a person has no legitimate expectation of privacy in information he voluntarily turns over to third parties."[84]

Miller and *Smith* are often cited as the Supreme Court's foundational "third-party doctrine" cases. For decades, this doctrine has maintained that when an individual voluntarily shares information with third parties, like telephone companies, banks, or even other individuals, the government can acquire that information from the third-party absent a warrant without violating the individual's Fourth Amendment rights. Law enforcement continues to rely on the third-party doctrine to obtain information that can be critical in criminal and national security investigations that keep the American people safe, and federal courts continue to apply the doctrine to both tangible and electronic information in a wide variety of contexts.

Against this backdrop, Congress and state legislatures have enacted statutes that provide additional safeguards for certain types of information, such as the Privacy Act of 1974 protecting personal information held by the federal government; the Electronic Communications Privacy Act of 1986 protecting (among other things) stored electronic communications; and the Pen/Trap Act protecting (among other things) dialing information for phone calls. These legislative measures provide statutory protection in the absence of a strong Fourth Amendment right to protect records held by third parties.

In light of technological advances, especially the creation of exponentially more electronic records about personal interactions, some commentators have called for a reexamination of third-party doctrine.[85] In 2010, the Sixth Circuit Court of Appeals in *United States v. Warshak* held that a subscriber has a reasonable expectation of privacy in his or her email communications, "analogous to a letter or a phone call" and that the government may not compel a commercial internet service provider to turn over the contents of a subscriber's emails without first obtaining a warrant based on probable cause.[86] In a recent Supreme Court case, Justice Sotomayor expressed the view in her concurring opinion that current practices around information disclosure to third parties are "ill-suited to the digital age, in which people reveal a great deal of information about themselves to third parties in the course of carrying out mundane tasks."[87]

Although we are not aware of any courts that have ruled that electronic content of communications can be accessed with less than a warrant, except with the consent of the user, since the *Warshak* case, the third-party doctrine has continued to apply to metadata of such communication and has been adapted and applied to cell-site location information and WiFi signals.[88]

This review of big data and privacy has cast even more light on the profound issues of privacy, market confidence, and rule of law raised by the manner in which the government compels the disclosure of electronic data. We will continually need to examine our laws and policy to keep pace with technology, and should consider how the protection of content data stored remotely, for instance with a cloud provider, should relate to the protection of content data stored in a home office or on a hard drive. This is true of emails, text messages, and other communications platforms, which over the past 30 years have become an important means of private personal correspondence, and are most often stored remotely.

Data and Metadata

The average American transacts with businesses in one form or another multiple times a day, from purchasing goods to uploading digital photos. These interactions create records, some of which, like pharmacy purchases, contain intimate personal information. In the course of ordinary activities, users also emit lots of "digital exhaust," or trace data, that leaves behind more fragmentary bits of information, such as the geographical coordinates of a cell phone transmission or an IP address in a server log. The advent of more powerful analytics, which can discern quite a bit from even small and disconnected pieces of data, raises the possibility that data gathered and held by third parties can be amalgamated and analyzed in ways that reveal even more information about individuals. What protections this material and the information derived from it merit is now a pressing question.

An equally profound question is whether certain types of data—specifically the "metadata" or transactions records about communications and documents, versus the content of those communications and documents—should be accorded stronger privacy protections than they are currently. "Metadata" is a term describing the character of the data itself. The classic example comes from telecommunications. The phone numbers originating and terminating a call, as metadata, are considered less revealing than the conversation itself and have been accorded different privacy protections.

Today, with the advent of big data, both the premise and policy may not always be so straightforward.

Experts seem divided on this issue, but those who argue that metadata today raises more sensitivities than in the past make a sufficiently compelling case to motivate review of policy on the matter. In the intelligence context, the President has already directed his Intelligence Advisory Board to consider the issue, and offer recommendations about the long-term viability of current assumptions about metadata and privacy. This review recommends that the government should broaden that examination beyond intelligence and consider the extent to which data and information should receive legal or other protections on the basis of how much it reveals about individuals.

Government Use of Commercial Data Services

Powerful private-sector profiling and data-mining technologies are not only used for commercial purposes. State, local, and federal agencies purchase access to many kinds of private databases for legitimate public uses, from land management to administering benefits. The sources of data that flow into these products are sometimes not publicly disclosed or may even be shielded as proprietary business information. Some legal scholars and privacy advocates have already raised concerns about the use of commercial data service products by the government, including law enforcement and intelligence agencies.[89]

The Department of the Treasury has been working to implement a program to help prevent waste, fraud, and abuse in federal spending by reducing the number of payments made to the wrong person, for the wrong amount, or without the proper paperwork. To provide federal agencies with a "one-stop-shop" to check various databases and identify ineligible recipients or prevent fraud or errors, the Treasury launched a "Do Not Pay" portal. While all of the current databases available on the portal are government databases, Treasury anticipates that commercial databases may eventually be useful as well.

To assist the Treasury, the Office of Management and Budget issued substantial guidance to ensure that individual privacy is fully protected in the program.[90] The guidance recognized that commercial data sources "may also present new or increased privacy risks, such as databases with inaccurate or out-of-date information." The guidelines require any commercial databases included in the Do Not Pay portal to be reviewed and approved following a 30-day period of public notice and comment. Among other requirements, the database must be relevant and necessary to the program, must be sufficiently accurate to ensure fairness to the individuals included in the database, and must not contain information that describes how any individual exercises rights guaranteed by the First Amendment, unless use of the data is expressly authorized by statute.

Given the increasing range of sensitive information available about individuals through commercial sources, this guidance is a significant step to ensure privacy protections when private-sector data is used to inform government decision-making. Similar OMB guidance should be considered for a wider range of agencies and programs, so the protections Americans have come to expect from their government exist regardless of where data originates.

Insider Threat and Continuous Evaluation

The 2013 shooting at the Washington Navy Yard facility by a contract employee who held a secret security clearance despite a record of arrests and troubling behavior has added urgency to ongoing efforts to more frequently evaluate employees who hold special positions of public trust.[91] It was the latest in a string of troubling breaches and acts of violence by insiders who held security clearances, including Chelsea Manning's disclosures to WikiLeaks, the Fort Hood shooting by Major Nidal Hasan, and the most serious breach in the history of U.S. intelligence, the release of classified National Security Agency documents by Edward Snowden.

Federal government employees and contractors go through different levels of investigation, depending on the level of risk, sensitivity of their position, or their need to access sensitive facilities or systems. Currently, employees and contractors who hold "top secret" clearances are reinvestigated every five years, and those holding "secret" clearances every ten. These lengthy gaps do not allow agencies to discover new and noteworthy information about an employee in a timely manner.

Pilot programs have demonstrated the efficacy of using automated queries of appropriate official and commercial databases and social media to identify violations or irregularities, known as "derogatory information," that may call into question a person's suitability to continue serving in a sensitive position. The Department of Defense, for instance, recently conducted a pilot of what it calls the "Automated Continuous Evaluation System." The pilot examined a sample of 3,370 Army service members, civilian employees, and contractor personnel, and identified that 21.7 percent of the tested population had previously unreported derogatory information that had developed since the last investigation. For 99 individuals, the pilot surfaced serious financial, domestic abuse, drug abuse, or allegations of prostitution that resulted in the revocation or suspension of their clearances.[92]

The Administration recently released a review of suitability and security practices which called for expanding continuous evaluation capabilities across the federal government.[93] The Administration's report recommends adopting practices across all agencies and security levels, although the exact extent of the information that will be used in these programs, especially social media sources, is still being determined.

These reforms will create a fundamentally different process for granting and maintaining security clearances that stands to enhance our security and safety. As the Administration works to expand the use of continuous evaluation across federal agencies, the privacy of employees and contractors will have to be carefully considered. The ability to refute or correct errant information that triggers reviews must be built into the process for appealing denials or revocations of clearance. We must ensure the big data analytics powering continuous evaluation are used in ways that protect the public as well as the civil liberties and privacy rights of those who serve on their behalf.

Conclusion

When wrestling with the vexing issues big data raises in the public sector, it can be easy to lose sight of the tremendous opportunities these technologies offer to improve public services, grow the economy, and improve the health and safety of our communities. These opportunities are real and must be kept at the center of the conversation about big data.

Big data holds enormous power to make the provision of services more efficient across the entire spectrum of government activity and to detect fraud, waste, and abuse at higher rates. Big data can also help create entirely new forms of value. New sources of precise data about weather patterns can provide meaningful scientific insights about climate change, while the ability to understand energy and natural resource use can lead to greater efficiency and reduce overall consumption. The movement, storage, and analysis of data all stands to grow more efficient and powerful. The Department of Energy, for instance, is working to develop computer memory and supercomputing frameworks that will in turn yield entire new classes of analytics tools, driving the big data revolution faster still.

There is virtually no part of government that does not stand serve citizens better. The big data revolution will take hold across the entire government, not merely in departments and agencies that already have missions involving science and technology. Those departments and agencies that have not historically made wide use of advanced data analytics have perhaps the most significant opportunity to harness big data to benefit the citizens they serve.

The power of big data does not stop at the federal level. It will be equally transformational for states and municipalities. Cities and towns have emerged as some of the most innovative users of big data to improve service delivery. The federal agencies and programs that provide grants and technical assistance to cities, towns, and counties should promote the use of these transformational municipal technologies to the greatest extent possible, replicating the successes pioneered by New York City's Office of Data Analytics and Chicago's Smart Data project.

Making big data work for the public good also takes people with skills that are in short supply and high demand. A recent assessment of the ability of the public and nonprofit sectors to attract and retain technical talent sounded a strong note of alarm.[94] Though there are many young technologists who care deeply about public service and would welcome the chance to work in government, private sector opportunities are so comparatively attractive that these technologists tend to use their skills applying big data in the marketplace rather than the public sector. This means that alongside investments in technology, the federal government must create a more attractive working culture for technologists and remove hiring barriers that keep out the very experts whose creativity and technical imagination is paramount to realizing the full potential of big data in government.

IV. PRIVATE SECTOR MANAGEMENT OF DATA

Big data means big things all across the global economy. In the next two years, the big data technologies and services market is projected to continue its rapid ascent.[95] This section considers how big data is shaping the products and services available to consumers and businesses, and highlights some of the challenges that arise when consumers have little insight into how information about them is being collected, analyzed, and used.

The Obama Administration has supported America's leadership position in using big data to spark innovation, productivity, and value in the private sector. However, the near-continuous collection, transfer, and re-purposing of information in a big data world also raises important questions about individual control over personal data and the risks of its use to exploit vulnerable populations. While big data will be a powerful engine for economic growth and innovation, there remains the potential for a disquieting asymmetry between consumers and the companies that control information about them.

Big Data Benefits for Enterprise and Consumer

Big data is creating value for both companies and consumers. The benefits of big data can be felt across a range of sectors, in both large and small firms, as access to data and the tools for processing it are further democratized. In large enterprises, there are several drivers of investment in big data technologies: the ability to analyze operational and transactional data, to glean insights into the behavior of online customers, to bring new and exceedingly complex products to market, and to derive deeper understanding from machines and devices within organizations.

Technology companies are using big data to analyze millions of voice samples to deliver more reliable and accurate voice interfaces. Banks are using big data techniques to improve fraud detection. Health care providers are leveraging more detailed data to improve patient treatment. Big data is being used by manufacturers to improve warranty management and equipment monitoring, as well as to optimize the logistics of getting their products to market. Retailers are harnessing a wide range of customer interactions, both online and offline, in order to provide more tailored recommendations and optimal pricing.[96]

For consumers, big data is fueling an expansion of products and services that impact their daily lives. It is enabling cybersecurity experts to protect systems—from credit card readers to electricity grids—by harnessing vast amounts of network and application data and using it to identify anomalies and threats.[97] It is also enabling some of the nearly 29 percent of Americans who are "unbanked" or "underbanked" to qualify for a line of credit by using a wider range of non-traditional information—such as rent payments, utilities, mobile-phone subscriptions, insurance, child care, and tuition—to establish creditworthiness.[98]

These new technologies are sensor-rich and embedded in networks. Lighting infrastructure can now detect sound, speed, temperature, and even carbon monoxide levels, and will draw data from car parks, schools, and along public streets to improve energy efficiency and public safety. Vehicles record and report a spectrum of driving and usage data that will pave the way for advanced transportation systems and improved safety. Home appliances can now tell us when to dim our lights from a thousand miles away. These are the kinds of changes that policies must accommodate. The Federal Trade Commission has already begun working to frame the policy questions raised by the Internet of Things, building on their long history of protecting consumers as new technologies come online.

The next sections discuss the online advertising and data services industries, each of which have significant histories using large datasets within long-established regulatory frameworks.

The Advertising-Supported Ecosystem

Since the earliest days of the commercial web, online advertising has been a vital driver of the growth of the Internet. One study estimated that the ad-supported Internet sustains millions of jobs in the United States and that the interactive marketing industry contributes billions to the U.S. economy each year.[99] This is a natural industry for big data to take root in and flourish. Increasingly precise data about consumers—where they are, what devices they use, and literally hundreds of categories of their interests—coupled with powerful analysis have enabled advertisers to more efficiently reach customers. Expensive television slots or full-page national magazine ads seem crude compared to the precisely segmented and instantaneously measured online ad marketplace. One study suggests that advertisers are willing to pay a premium of between 60 and 200 percent for online targeted advertising.[100]

Consumers are reaping the benefits of a robust digital ecosystem that offers a broad array of free content, products, and services. The Internet also puts national or international advertising within reach of not just major companies, but mom-and-pop stores and fledgling brands. As a result, consumers are getting better, more useful ads from—and access to—a wider range of businesses, in a marketplace that is ultimately more competitive and innovative.

Many different actors play a role in making this ecosystem work, including the consumer, the companies they engage with directly, and an array of other entities that provide services like analytics or security, or derive and share data. Standing between the publisher of the website a user visits and the advertiser paying for the ad displayed on the user's page are a dizzying array of other companies. Advertising networks and ad exchanges facilitate transactions between the publishers and the advertisers. Ad content and campaigns are created and placed by agencies, optimizers, and media planners.

Ad performance is measured and analyzed by yet another set of specialized companies.[101]

In general, the companies with which a consumer engages directly—news websites, social media, or online or offline retailers—are called "first parties," as they collect information directly from the consumer. But as described above, a broad range of companies may gather information indirectly because they are in the business of processing data on behalf of the first-party company or may have access to data—most often in an aggregated or de-identified form—as part of a different business relationship. These "third-party" companies include the many "middle players" in the digital ecosystem, as well as financial transaction companies that handle payment processing, companies that fill orders, and others. The first parties may use the data themselves, or resell it to others to develop advertising profiles or for other uses. Users, more often than not, do not understand the degree to which they are a commodity in each level of this marketplace.

The Consumer and the Challenge of Transparency

For well over a decade, the online advertising industry has worked to provide consumers choice and transparency in a self-regulatory framework. Starting at the edges of the ecosystem, where the consumer can identify the website publisher and the advertiser whose ads are served, privacy policies and other forms of notice have served to inform consumers how their information is used. Under this self-regulatory regime, companies agree to a set of principles when engaged in "behavioral" or multi-site advertising where they collect information about user activities over time and across different websites in order to infer user preferences. These principles include requiring notice to the user about their data collection practices; providing options for users to opt out of some forms of tracking; limiting the use of sensitive information, such as children's information or medical or financial data; and a requirement to delete or de-identify data.

Technologies to improve transparency and privacy choices online have been slow to develop, and for many reasons have not been used widely by consumers. For example, under the self-regulatory regime adopted by advertisers and ad networks, many online behavioral ads include a standardized icon that indicates information is being collected for purposes of behavioral ad targeting, and links to a page where the consumer can opt-out of such collection.[102] According to the online advertising industry, this icon has appeared on ads billions of times, but only a tiny fraction of users utilize this feature or understand its meaning. Advertising networks operated by some of the largest online companies have also offered users detailed dashboards for seeing the basis on which they are targeted for advertising and giving them the ability to opt out.[103] These, too, have received little consumer attention. There are many theories about why users do not make use of these privacy features. Some assert that the privacy tools are hidden or too difficult for most users to navigate.[104] Others argue that users have "privacy fatigue" from the barrage of privacy policies and settings they must wade through to simply use a service.[105] It is also possible that most of the public is not very bothered by personalized ads when they enjoy a robust selection of free content, products, and services.

As we look ahead at the rising trajectory of information collection across many sources and the ability to target advertising with greater precision, the challenge to consumer transparency and meaningful choice deepens. Even employing relatively straightforward technical measures that would provide consumers with greater control over how data flows between their web browser and the servers of the webpages they visit for advertising purposes—what has become known as the "Do Not Track" browser setting—can be problematic because anti-fraud and online security activities now rely on these same data flows to track and prevent malicious activity.

THE CHALLENGE WITH DO NOT TRACK

The idea behind a Do Not Track privacy setting is to provide an easy-to-use solution that empowers consumers to limit the tracking of their activities across websites. Some browsers provide a kind of Do Not Track capability by blocking third-party cookies by default, or allowing consumers to choose to do so. Some browsers also allow consumers to send a signal instructing services not to track them. While Do Not Track technology is fairly straightforward, attempts to build consensus around the policy requirements for the websites receiving visits by users with Do Not Track technology enabled have proven far more difficult.

Some websites voluntarily agreed to honor the wishes of visitors with Do Not Track indicators, but others have not, or have adopted policies that still permit partial tracking—muddling expectations for consumers and frustrating privacy advocates.

> A working group of the World Wide Web Consortium, which included technologists, developers, advertising industry representatives, and privacy advocates, worked to craft a standard for implementation of the Do Not Track signal for more than three years. Recently, the working group released a final candidate for a technical Do Not Track specification, which will now go to the larger community to consider for approval.
>
> In the meantime, the European Union amended its E-Privacy Directive in 2009 to require user consent to the use of cookies and other online tracking devices, unless they are "strictly necessary for delivery of a service requested by the user," such as an online shopping cart. Compliance with the Directive has been uneven, although many European company websites now obtain a one-time explicit consent for the use of cookies—a solution that is widely acknowledged as clunky and which has been criticized in some circles as not providing the user the meaningful choice about privacy first envisioned by the directive.
>
> While imperfect, these efforts reflect a growing interest in creating a technological means to allow individuals to control how commercial entities collect and use information about them.

The Data Services Sector

Alongside firms that focus primarily on online advertising are a related set of businesses that offer broader services drawn from information about consumers, public records, and other data sets. The "data services" sector—sometimes called "data brokers"— encompasses a class of businesses that collect data across many sources, aggregate and analyze it, and then share that information, or information derived from it. Typically, these companies have no direct relationship with the consumers whose information they collect. Instead, they offer services to other businesses or government agencies, including marketing products, verifying an individual's identity, providing "people search" services, or detecting fraud. Some of these companies also have a specific line of business as "consumer reporting agencies," which provide reports for purposes of credit applications, insurance, employment, or health care reports.

From a regulatory standpoint, data services fall into three broad categories:

1. Consumer reporting functions regulated under the Fair Credit Reporting Act, which generally keep the data, analysis, and reporting collected and used for these purposes in a separate system and under specific compliance rules apart from the rest of their data services operations.
2. Risk mitigation services such as identity verification, fraud detection and people-search or look up services; and
3. Marketing services to identify potential customers, enhance ad targeting information, and other advertising-related services.

The Fair Credit Reporting Act, as discussed in Section 2, provides affirmative rights to consumers. Consumer reporting agencies that provide reports for determining eligibility for credit, insurance, or employment, are required under the Fair Credit Reporting Act or the Equal Credit Opportunity Act to inform consumers when an adverse action, such as a denial or higher cost of credit, is taken against them based on a report. By law, consumers also have a right to know what is in their file, what their credit score is, and how to correct or delete inaccurate information.[106] The Fair Credit Reporting Act mandates that credit reporting agencies remove negative information after certain periods, such that late payments and tax liens are deleted from a consumer's file after seven years and bankruptcies after ten. Certain types of information—such as race, gender and religion—may not be used as factors to determine creditworthiness.

These statutory rights do not exist for risk mitigation or marketing services. As a matter of practice, data services companies may provide access and correction mechanisms to consumers for the information used in identity verification. In the context of marketing services, some companies permit consumers to opt-out of having their personal information used in marketing services.

Unregulated Data Broker Services

To assist marketers, data brokers can provide a profile of a consumer who may interact with a brand or seek services across many different channels, from online web presence to social media to mobile engagement. Data brokers aggregate purchase patterns, activities on a website, mobile, social media, ad network interactions, or direct customer support, and then further "enhance" it with information from public records or other commercially available sources. That information is used to develop a profile of a customer, whose activities or engagements can then be monitored to help the marketer pinpoint the message to send and the right moment to send it.

These profiles can be exceptionally detailed, containing upwards of thousands of pieces of data. Some large data firms have profiles on hundreds of millions of consumers. They algorithmically analyze this information to segment customers into precise categories, often with illustrative names that help their business customers identify populations for targeted advertising. Some of these categories include "Ethnic Second-City Strugglers," "Retiring on Empty: Singles," "Tough Start: Young Single Parents," "Credit Crunched: City Families," and "Rural and Barely Making It."[107] These products include factual information about individuals as well as "modeled" elements inferred from other data. Data brokers then sell "original lists" of consumers who fit particular criteria. They may also offer a "data append" service whereby companies can buy additional data about particular customers to help them build out more complete profiles of individuals on whom they maintain information.[108]

WHAT IS A CREDIT REPORTING AGENCY?

Since the 1950s, credit reporting companies–now known as "consumer reporting agencies"–have collected information and provided reports on individuals that are used to decide eligibility for credit, insurance or a job. In one typical scenario, a credit reporting agency collects information about an individual's credit history, such as whether they pay their bills on time, how many and what kind of accounts they hold and for how long, whether they've been the subject of collection actions, and whether they have outstanding debt. The agency then uses a statistical program to compare this information to the loan repayment history of consumers with similar profiles and assigns a score that reflects the individual's creditworthiness: how likely it is that he or she will repay a loan and make timely payments. This score facilitates consumers' ability to buy a home or car or otherwise engage in the economy by becoming a basis for creditors' decisions about whether to provide credit to the consumer, and on what terms.

While this precise profiling of consumer attributes yields benefits, it also represents a powerful capacity on the part of the private sector to collect information and use that information to algorithmically profile an individual, possibly without the individual's knowledge or consent. This application of big data technology, if used improperly, irresponsibly, or nefariously, could have significant ramifications for targeted individuals. In its 2012 Privacy Report, the Federal Trade Commission recommended that data brokers become more transparent in the services that are not already covered by the Fair Credit Report Act, and provide consumers with reasonable access to and choices about data maintained about them, in proportion to the sensitivity of data and how it is used.[109]

Algorithms, Alternative Scoring and the Specter of Discrimination

The business models and big data strategies now being built around the collection and use of consumer data, particularly among the "third-party" data services companies, raise important questions about how to ensure transparency and accountability in these practices. Powerful algorithms can unlock value in the vast troves of information available to businesses, and can help empower consumers, but also raise the potential of encoding discrimination in automated decisions. Fueled by greater access to data and powerful analytics, there are now a host of products that "score" individuals beyond the scope of traditional credit scores, which are regulated by law.[110] These products attempt to statistically characterize everything from a consumer's ability to pay to whether, on the basis of their social media posts, they are a "social influencer" or "socially influenced."

While these scores may be generated for marketing purposes, they can also in practice be used similarly to regulated credit scores in ways that influence an individuals' opportunities to find housing, forecast their job security, or estimate their health, outside of the protections of the Fair Credit Reporting Act or Equal Credit Opportunity Act.[111] Details on what types of data are included in these scores and the algorithms used for assigning attributes to an individual are held closely by companies and largely invisible to consumers. That means there is often no meaningful avenue for either identifying harms or holding any entity in the decision-making chain accountable.

Because of this lack of transparency and accountability, individuals have little recourse to understand or contest the information that has been gathered about them or what that data, after analysis, suggests.[112] Nor is there an industry-wide portal for consumers to communicate with data services companies, as the online advertising industry voluntarily provides and the Fair Credit Reporting Act requires for regulated entities. This can be particularly harmful to victims of identity theft who have ongoing errors or omissions impacting their scores and, as a result, their ability to engage in commerce.

> ## WHAT IS AN ALGORITHM?
>
> In simple terms, an algorithm is defined by a sequence of steps and instructions that can be applied to data. Algorithms generate categories for filtering information, operate on data, look for patterns and relationships, or generally assist in the analysis of information. The steps taken by an algorithm are informed by the author's knowledge, motives, biases, and desired outcomes. The output of an algorithm may not reveal any of those elements, nor may it reveal the probability of a mistaken outcome, arbitrary choice, or the degree of uncertainty in the judgment it produces. So-called "learning algorithms" which underpin everything from recommendation engines to content filters evolve with the datasets that run through them, assigning different weights to each variable. The final computer-generated product or decision—used for everything from predicting behavior to denying opportunity—can mask prejudices while maintaining a patina of scientific objectivity.

For all of these reasons, the civil rights community is concerned that such algorithmic decisions raise the specter of "redlining" in the digital economy—the potential to discriminate against the most vulnerable classes of our society under the guise of neutral algorithms.[113] Recently, some offline retailers were found to be using an algorithm that generated different discounts for the same product to people based on where they believed the customer was located. While it may be that the price differences were driven by the lack of competition in certain neighborhoods, in practice, people in higher-income areas received higher discounts than people in lower-income areas.[114]

There are perfectly legitimate reasons to offer different prices for the same products in different places. But the ability to segment the population and to stratify consumer experiences so seamlessly as to be almost undetectable demands greater review, especially when it comes to the practice of differential pricing and other potentially discriminatory practices. It will also be important to examine how algorithmically-driven decisions might exacerbate existing socio-economic disparities beyond the pricing of goods and services, including in education and workforce settings.

Conclusion

The advertising-supported Internet creates enormous value for consumers by providing access to useful services, news, and entertainment at no financial cost. The ability to more precisely target advertisements is of enormous value to companies, which can efficiently reach audiences that are more likely to purchase their goods and services. However, private-sector uses of big data must ensure vulnerable classes are not unfairly targeted. The increasing use of algorithms to make eligibility decisions must be carefully monitored for potential discriminatory outcomes for disadvantaged groups, even absent discriminatory intent. The Federal Trade Commission should be commended for their continued engagement with industry and the public on this complex topic and should continue its plans to focus further attention on emerging practices in the data broker industry. We look forward to their forthcoming report on this important topic. Additional work should be done to identify practical ways of increasing consumer access to information about unregulated consumer scoring, with particular emphasis on the ability to correct or suppress inaccurate information. Likewise, additional research in measuring adverse outcomes due to the use of scores or algorithms is needed to understand the impacts these tools are having and will have in both the private and public sector as their use grows.

V. TOWARD A POLICY FRAMEWORK FOR BIG DATA

In what feels like the blink of an eye, the information age has fundamentally reconfigured how data affects individual lives and the broader economy. More than 6,000 data centers dot the globe. International data flows are continuous and multidirectional. To a greater degree than ever before, this data is being harnessed by businesses, governments, and entrepreneurs to improve the services they deliver and enhance how people live and work.

Big data applications create social and economic value on a scale that, collectively, is of strategic importance for the nation. Technological innovation is the animating force of the American economy. In the years to come, big data will foster significant productivity gains in industry and manufacturing, further accelerating the integration of the industrial and information economies.

Government should support the development of big data technologies with the full suite of policy instruments in its toolkit. Agencies must continue advancing the Administration's Open Data initiative. The federal government should also invest in research and development to support big data technologies, especially as they apply to education, health care, and energy. As the preceding sections have documented, adjusting existing policies will make possible certain new applications of big data that are clearly in the public interest, particularly in health care. The policy framework for big data will require cooperation between the public and private sectors to accelerate the revolution that is underway and identify barriers that ought to be removed for innovations driven by big data to flourish.

Like other transformative factors of production, big data generates value differently for individuals, organizations, and society. While many applications of big data are unequivocally beneficial, some of its uses impact privacy and other core values of fairness, equity, and autonomy.

Big data technologies enable data collection that is more ubiquitous, invasive, and valuable. This new cache of collected and derived data is of huge potential benefit but is also unevenly regulated. Certain private and public institutions have access to more data and more resources to compute it, potentially heightening asymmetries between institutions and individuals.

It is the responsibility of government to ensure that transformative technologies are used fairly and employed in all areas where they can achieve public good. Four areas in particular emerge as places for further policy exploration:

1. How government can harness big data for the public good while guarding against unacceptable uses against citizens;
2. The extent to which big data alters the consumer landscape in ways that implicate core values;
3. How to protect citizens from new forms of discrimination that may be enabled by big data technologies; and
4. How big data affects the core tenet of modern privacy protection, the notice and consent framework that has been in wide use since the 1970s.

Big Data and the Citizen

Big data will enhance how the government administers public services and enable it to create whole new kinds of value. But big data tools also unquestionably increase the potential of government power to accrue unchecked. Local police departments now have access to surveillance tools more powerful than those used by superpowers during the Cold War. The new means of surveillance that in Justice Alito's evocative analogy deploy "tiny constables" to all areas of life, together with the ways citizens can be profiled by algorithms that redirect police powers, raise many questions about big data's implications for First Amendment rights of free speech and free association.

Many of the laws governing law enforcement access to electronic information were passed by Congress at a time when private papers were largely stored in the home. The Stored Communications Act, which is part of the Electronic Communications Privacy Act (ECPA), articulates the rules for obtaining the content of electronic communications, including email and cloud services. ECPA was originally passed in 1986. It has served to protect the privacy of individuals' stored communications. But with time, some of the lines drawn by the statute have become outdated and no longer reflect ways in which we use technology today. In considering how to update the Act, there are a variety of interests at stake, including privacy interests and the need for law enforcement and civil enforcement agencies to protect public safety and enforce criminal and civil law. Email, text messaging, and other private digital communications have become the principal means of personal correspondence and the cloud is increasingly used to store individuals' files. They should receive commensurate protections.

Similarly, many protections afforded to metadata were calibrated for a time that predated the rise of personal computers, the Internet, mobile phones, and cloud computing. No one imagined then that the traces of digital data left today as a matter of routine can be reassembled to reveal intimate personal details. Today, most law enforcement uses of metadata are still rooted in the "small data" world, such as identifying phone numbers called by a criminal suspect. In the future, metadata that is part of the "big data" world will be increasingly relevant to investigations, raising the question of what protections it should be granted. While today, the content of communications, whether written or verbal, generally receives a high level of legal protection, the level of protection afforded to metadata is less so.

Although the use of big data technologies by the government raises profound issues of how government power should be regulated, big data technologies also hold within them solutions that can enhance accountability, privacy, and the rights of citizens. These include sophisticated methods of tagging data by the authorities under which it was collected or generated; purpose-and user-based access restrictions on this data; tracking which users access what data for what purpose; and algorithms that alert supervisors to possible abuses. All of these methods are being employed in parts of the federal government today to protect the rights of citizens and regulate how big data technologies are used, and more agencies should put them to use.

Responsibly employed, big data could lead to an aggregate increase in actual protections for the civil liberties and civil rights afforded of citizens, as well as drive transformation improvements in the provision of public services.

Big Data and the Consumer

The technologies of collection and analysis that fuel big data are being used in every sector of society and the economy. Many of them are trained squarely on people as consumers. One of the most intensely discussed of big data analytics to date has been in the online advertising industry, where it is used to serve customized ads as people browse the web or travel around town with their mobile phone. But the information collected and the uses to which it is put are far broader and quickly changing, with data derived from the real world increasingly being combined with data drawn from online activity.

The end result is a massive increase in the amount of intimate information compiled about individuals. This information is highly valuable to businesses of all kinds. It is bought, bartered, traded, and sold. An entire industry now exists to commoditize the conclusions drawn from that data. Products sold on the market today include dozens of consumer scores on particular individuals that describe attributes, propensities, degrees of social influence over others, financial habits, household wealth, and even suitability as a tenant, job security, and frailty. While some of these scoring efforts are highly regulated, other uses of data are not.

There are enormous benefits associated with the rise of profiling and targeted advertising and the ways consumers can be tracked and offered services as they move through the online and physical world. Advertising and marketing effectively subsidize many free goods on the Internet, fueling an entire industry in software and consumer apps. As one person pointedly remarked during this review, "We don't like putting a quarter into the machine to go do a web search."

Data collection is also vital to securely verify identity online. The data services and financial industries have gone to extraordinary lengths to enable individuals to conduct secure transactions from computers and mobile devices. The same verification technologies that make transaction in the private sector possible also enable citizens to securely interact with the government online, opening a new universe of public services, all accessible from an arm chair.

But there are also costs to organizing the provision of commercial services in this way. Amalgamating so much information about consumers makes data breaches more consequential, highlighting the need for federal data breach legislation to replace a confusing patchwork of state standards. The sheer number of participants in this new, interconnected ecosystem of data collection, storage, aggregation, transfer, and sale can disadvantage consumers. The average consumer is unlikely to be aware of the range of data being collected or held or even to know who holds it; will have few opportunities to engage over the scope or accuracy of data being held about them; and may have limited insight into how this information feeds into algorithms that make decisions about their consumer experience or market access.

When considering what policies will allow big data to flourish in the consumer context, a crucial distinction must be drawn around the ways this collected information gets used. It is one thing for big data to segment consumers for marketing purposes, thereby providing more tailored opportunities to purchase goods and services. It is another, arguably far more serious, matter if this information comes to figure in decisions about a consumer's eligibility for—or the conditions for the provision of—employment, housing, health care, credit, or education.

Big Data and Discrimination

In addition to creating tremendous social good, big data in the hands of government and the private sector can cause many kinds of harms. These harms range from tangible and material harms, such as financial loss, to less tangible harms, such as intrusion into private life and reputational damage. An important conclusion of this study is that big data technologies can cause societal harms beyond damages to privacy, such as discrimination against individuals and groups. This discrimination can be the inadvertent outcome of the way big data technologies are structured and used. It can also be the result of intent to prey on vulnerable classes.

An illustrative example of how one organization ensured that a big data technology did not inadvertently discriminate comes from Boston, where the city developed an experimental app in partnership with the Mayor's Office of New Urban Mechanics.[115] Street Bump is a mobile application that uses a smartphone's accelerometer and GPS feed to collect data about road condition, including potholes, and report them to the city's Public Works Department. It is a marvelous example of how cities are creatively using crowdsourcing to improve service delivery. But the Street Bump team also identified a potential problem with deploying the app to the public. Because the poor and the elderly are less likely to carry smartphones or download the Street Bump app, its release could have the effect of systematically directing city services to wealthier neighborhoods populated by smartphone owners.

To its credit, the city of Boston and the StreetBump developers figured this out before launching the app. They first deployed it to city-road inspectors, who service all parts of the city equally; the public now provides additional supporting data. It took foresight to prevent an unequal outcome, and the results were worth it. The Street Bump app has to date recorded 36,992 "bumps," helping Boston identify road castings like manholes and utility covers, not potholes, as the biggest obstacle for drivers.

More serious cases of potential discrimination occur when individuals interact with complex databases as they verify their identity. People who have multiple surnames and women who change their names when they marry typically encounter higher rates of error. This has also been true, for example, in the E-verify program, a database run jointly by the Department of Homeland Security and the Social Security Administration, which has long been a concern for civil rights advocates.

E-verify provides employers the ability to confirm the eligibility of newly hired employees to work legally in the United States. Especially given the number of queries the system processes and the volume of information it amalgamates from different sources that are themselves constantly changing, the overwhelming majority of results returned by E-verify are timely and accurate, giving employers certainty that people they hire are authorized to work in the United States. Periodic evaluations to improve the performance of E-verify have nonetheless revealed different groups receive initial verifications at different rates. A 2009 evaluation found the rate at which U.S. citizen have their authorization to work be initially erroneously unconfirmed by the system was 0.3 percent, compared to 2.1 percent for non-citizens. However, after a few days many of these workers' status was confirmed.[116]

The Department of Homeland Security and Social Security Administration have focused great attention on addressing this issue. A more recent evaluation of the program found many more people were able to verify their work status more quickly and with lower rates of error. Over five years, the rates of initial mismatch fell by 60 percent for U.S. citizens and 30 percent for non-citizens.[117] Left unresolved, technical issues like this could create higher barriers to employment or other critical needs for certain individuals and groups, making imperative the importance of accuracy, transparency, and redress in big data systems.

These two examples of inadvertent discrimination illustrate why it is important to monitor outcomes when big data technologies are applied even in instances where discriminatory intent is not present and where one might not anticipate an inequitable impact. There is, however, a whole other class that merits concern—the use of big data for deliberate discrimination.

We have taken considerable steps as a society to mandate fairness in specific domains, including employment, credit, insurance, health, housing, and education. Existing legislative and regulatory protections govern how personal data can be used in each of these contexts. Though predictive algorithms are permitted to be used in certain ways, the data that goes into them and the decisions made with their assistance are subject to some degree of transparency, correction, and means of redress. For important decisions like employment, credit, and insurance, consumers have a right to learn why a decision was made against them and what information was used to make it, and to correct the underlying information if it is in error.

These protections exist because of the United States' long history of discrimination. Since the early 20th century, banks and lenders have used location data to make assumptions about individuals. It was not until the Home Mortgage Disclosure Act was signed into law in 1975 that denying granting a person a loan on the basis of what neighborhood they live in rather than their personal capacity for credit became far less prevalent. "Redlining," in which banks quite literally drew—and in cases continue to draw—boundaries around neighborhoods where they would not loan money, existed for decades as a potent tool of discrimination against African-Americans, Latinos, Asians, and Jews.

Just as neighborhoods can serve as a proxy for racial or ethnic identity, there are new worries that big data technologies could be used to "digitally redline" unwanted groups, either as customers, employees, tenants, or recipients of credit. A significant finding of this report is that big data could enable new forms of discrimination and predatory practices.

The same algorithmic and data mining technologies that enable discrimination could also help groups enforce their rights by identifying and empirically confirming instances of discrimination and characterizing the harms they caused. Civil rights groups can use the new and powerful tools of big data in service of equal treatment for the communities they represent. Whether big data will build greater equality for all Americans or exacerbate existing inequalities depends entirely on how its technologies are applied in the years to come, what kinds of protections are present in the law, and how the law is enforced.

Big Data and Privacy

Big data technologies, together with the sensors that ride on the "Internet of Things," pierce many spaces that were previously private. Signals from home WiFi networks reveal how many people are in a room and where they are seated. Power consumption data collected from demand-response systems show when you move about your house.[118] Facial recognition technologies can identify you in pictures online and as soon as you step outside. Always-on wearable technologies with voice and video interfaces and the arrival of whole classes of networked devices will only expand information collection still further. This sea of ubiquitous sensors, each of which has legitimate uses, make the notion of limiting information collection challenging, if not impossible.

This trend toward ubiquitous collection is in part driven by the nature of technology itself.[119] Whether born analog or digital, data is being reused and combined with other data in ways never before thought possible, including for uses that go beyond the intent motivating initial collection. The potential future value of data is driving a digital land grab, shifting the priorities of organizations to collect and harness as much data as possible. Companies are now constantly looking at what kind of data they have and what data they need in order to maximize their market position. In a world where the cost of data storage has plummeted and future innovation remains unpredictable, the logic of collecting as much data as possible is strong.

Another reality of big data is that once data is collected, it can be very difficult to keep anonymous. While there are promising research efforts underway to obscure personally identifiable information within large data sets, far more advanced efforts are presently in use to re-identify seemingly "anonymous" data. Collective investment in the capability to fuse data is many times greater than investment in technologies that will enhance privacy.

Together, these trends may require us to look closely at the notice and consent framework that has been a central pillar of how privacy practices have been organized for more than four decades. In a technological context of structural over-collection, in which re-identification is becoming more powerful than de-identification, focusing on controlling the collection and retention of personal data, while important, may no longer be sufficient to protect personal privacy. In the words of the President's Council of Advisors for Science & Technology, "The notice and consent is defeated by exactly the positive benefits that big data enables: new, non-obvious, unexpectedly powerful uses of data."[120]

Anticipating the Big Data Revolution's Next Chapter

For the vast majority of today's ordinary interactions between consumers and first parties, the notice and consent framework adequately safeguards privacy protections. But as the President's Council of Advisors on Science & Technology note, the trajectory of technology is shifting to far more collection, use and storage of data by entities that do not have a direct relationship with the consumer or individual.[122] In instances where the notice and consent framework threatens to be overcome—such as the collection of ambient data by our household appliances—we may need to re-focus our attention on the context of data use, a policy shift presently being debated by privacy scholars and technologists.[123] The context of data use matters tremendously.

Data that is socially beneficial in one scenario can cause significant harm in another. To borrow a term, data itself is "dual use." It can be used for good or for ill.

Putting greater emphasis on a responsible use framework has many potential advantages. It shifts the responsibility from the individual, who is not well equipped to understand or contest consent notices as they are currently structured in the marketplace, to the entities that collect, maintain, and use data. Focusing on responsible use also holds data collectors and users accountable for how they manage the data and any harms it causes, rather than narrowly defining their responsibility to whether they properly obtained consent at the time of collection.

Focusing more attention on responsible use does not mean ignoring the context of collection. Part of using data responsibly could mean respecting the circumstances of its original collection. There could, in effect, be a "no surprises" rule, as articulated in the "respect for context" principle in the Consumer Privacy Bill of Rights. Data collected in a consumer context could not suddenly be used in an employment one.

Technological developments support this shift toward a focus on use. Advanced data-tagging schemes can encode details about the context of collection and uses of the data already granted by the user, so that information about permissive uses travels along with the data wherever it goes. If well developed and brought widely into use, such a data-tagging scheme would not solve all the dilemmas posed by big data, but it could help address several important challenges.

FEDERAL RESEARCH IN PRIVACY-ENHANCING TECHNOLOGIES

The research and development of privacy enhancing technologies has been a priority for the Obama Administration. Agencies across the Networking and Information Technology Research and Development (NITRD) program collectively spend over $70 million each year on privacy research.[121]

This research falls into four broad areas: support for privacy as an extension of security; research on how enterprises comply with privacy laws; privacy in health care; and basic research into technologies that enable privacy. The table below summarizes some of the research programs in progress at agencies in the NITRD.

In their review of big data technologies, the President's Council of Advisors on Science & Technology endorses strengthening U.S. research in privacy-related technologies and the social science questions surrounding their use.

Research areas	Support for privacy as an extension of security	Research on how enterprises comply with privacy laws	Privacy in health care	Privacy research explorations
Agencies	Air Force Research Laboratory, Defense Advanced Research Projects Agency, National Security Agency, Intelligence Advanced Research Projects Activity, Office of Naval Research	Department of Energy, Department of Homeland Security, National Institute of Standards and Technology	Telemedicine and Advanced Technology Research Center, Office of the National Coordinator for Health Information Technology, National Institute of Health	National Science Foundation
Funding est. (total $77M/year)	$34M/year	$10M/year	$8M/year	$25M/year
Sampling of key projects	Anonymization techniques	Automated privacy compliance	Collection and use limitation	Algorithmic foundations for privacy and tools
	Confidential collaboration and communication	Location-privacy tools	Data segmentation for privacy	Economics of privacy

Research areas	Support for privacy as an extension of security	Research on how enterprises comply with privacy laws	Privacy in health care	Privacy research explorations
	Homomorphic encryption	Protection of personally identifiable information	Patient consent and privacy	Privacy as a social-psychological construct
	Homomorphic encryption	Standards for legal compliance	Patient data quality	Privacy policy analysis
	Traffic-secure routing	Voluntary code of conduct for smart grid	Preserving anonymity in health care data	Privacy solutions for cloud computing, data integration, mining

Perhaps most important of all, a shift to focus on responsible uses in the big data context allows us to put our attention more squarely on the hard questions we must reckon with: how to balance the socially beneficial uses of big data with the harms to privacy and other values that can result in a world where more data is inevitably collected about more things. Should there be an agreed-upon taxonomy that distinguishes information that you do not collect or use under any circumstances, information that you can collect or use without obtaining consent, and information that you collect and use only with consent? How should this taxonomy be different for a medical researcher trying to cure cancer and a marketer targeting ads for consumer products?

As President Obama said upon the release of the Consumer Privacy Bill of Rights, "Even though we live in a world in which we share personal information more freely than in the past, we must reject the conclusion that privacy is an outmoded value." Privacy, the President said, "has been at the heart of our democracy from its inception, and we need it now more than ever." This is even truer in a world powered by big data.

CONCLUSION AND RECOMMENDATIONS

The White House review of big data and privacy, announced by President Obama on January 17, 2014, was conceived to examine the broader implications of big data technology.

The President recognized the big data revolution is playing out widely across the public and private sectors and that its implications need to be considered alongside the Administration's review of signals intelligence.

The White House big data working group set out to learn, in 90 days, how big data technologies are transforming government, commerce, and society. We wanted to understand what opportunities big data affords us, and the advances it can spur.

We wanted a better grasp of what kinds of technologies already existed, and what we could anticipate coming just over the horizon. The President's Council of Advisors for Science & Technology conducted a parallel report to take measure of the underlying technologies. Their findings underpin many of the technological assertions in this report.

Big data tools offer astonishing and powerful opportunities to unlock previously inaccessible insights from new and existing data sets. Big data can fuel developments and discoveries in health care and education, in agriculture and energy use, and in how businesses organize their supply chains and monitor their equipment.

Big data holds the potential to streamline the provision of public services, increase the efficient use of taxpayer dollars at every level of government, and substantially strengthen national security. The promise of big data requires government data be viewed as a national resource and be responsibly made available to those who can derive social value from it. It also presents the opportunity to shape the next generation of computational tools and technologies that will in turn drive further innovation.

Big data also introduces many quandaries. By their very nature, many of the sensor technologies deployed on our phones and in our homes, offices, and on lampposts and rooftops across our cities are collecting more and more information. Continuing advances in analytics provide incentives to collect as much data as possible not only for today's uses but also for potential later uses.

Technologically speaking, this is driving data collection to become functionally ubiquitous and permanent, allowing the digital traces we leave behind to be collected, analyzed, and assembled to reveal a surprising number of things about ourselves and our lives. These developments challenge longstanding notions of privacy and raise questions about the "notice and consent" framework, by which a user gives initial permission for their data to be collected. But these trends need not prevent creating ways for people to participate in the treatment and management of their information.

An important finding of this review is that while big data can be used for great social good, it can also be used in ways that perpetuate social harms or render outcomes that have inequitable impacts, even when discrimination is not intended. Small biases have the potential to become cumulative, affecting a wide range of outcomes for certain disadvantaged groups. Society must take steps to guard against these potential harms by ensuring power is appropriately balanced between individuals and institutions, whether between citizen and government, consumer and firm, or employee and business.

The big data revolution is in its earliest stages. We will be grappling for many years to understand the full sweep of its technologies; the ways it will empower health, education, and the economy; and, crucially, what its implications are for core American values, including privacy, fairness, non-discrimination, and self-determination.

Even at this early juncture, the authors of this report believe important conclusions are already emerging about big data that can inform how the Administration moves forward in a number of areas. In particular, there are five areas that will each bring the American people into the national conversation about how to maximize benefits and minimize harms in a big data world:

1. **Preserving Privacy Values:** Maintaining our privacy values by protecting personal information in the marketplace, both in the United States and through interoperable global privacy frameworks;
2. **Educating Robustly and Responsibly:** Recognizing schools—particularly K-12—as an important sphere for using big data to enhance learning opportunities, while protecting personal data usage and building digital literacy and skills;
3. **Big Data and Discrimination:** Preventing new modes of discrimination that some uses of big data may enable;
4. **Law Enforcement and Security:** Ensuring big data's responsible use in law enforcement, public safety, and national security; and
5. **Data As a Public Resource:** Harnessing data as a public resource, using it to improve the delivery of public services, and investing in research and technology that will further power the big data revolution.

POLICY RECOMMENDATIONS

This review also identifies six discrete policy recommendations that deserve prompt Administration attention and policy development. These are:

- **Advance the Consumer Privacy Bill of Rights.** The Department of Commerce should take appropriate consultative steps to seek stakeholder and public comment on big data developments and how they impact the Consumer Privacy Bill of Rights and then devise draft legislative text for consideration by stakeholders and submission by the President to Congress.
- **Pass National Data Breach Legislation.** Congress should pass legislation that provides for a single national data breach standard along the lines of the Administration's May 2011 Cybersecurity legislative proposal.
- **Extend Privacy Protections to Non-U.S. Persons.** The Office of Management and Budget should work with departments and agencies to apply the Privacy Act of 1974 to non-U.S. persons where practicable, or to establish alternative privacy policies that apply appropriate and meaningful protections to personal information regardless of a person's nationality.
- **Ensure Data Collected on Students in School is Used for Educational Purposes.** The federal government must ensure that privacy regulations protect students against having their data being shared or used inappropriately, especially when the data is gathered in an educational context.
- **Expand Technical Expertise to Stop Discrimination.** The federal government's lead civil rights and consumer protection agencies should expand their technical expertise to be able to identify practices and outcomes facilitated by big data analytics that have a discriminatory impact on protected classes, and develop a plan for investigating and resolving violations of law.
- **Amend the Electronic Communications Privacy Act.** Congress should amend ECPA to ensure the standard of protection for online, digital content is consistent with that afforded in the physical world—including by removing archaic distinctions between email left unread or over a certain age.

1. Preserving Privacy Values

Big data technologies are driving enormous innovation while raising novel privacy implications that extend far beyond the present focus on online advertising. These implications make urgent a broader national examination of the future of privacy protections, including the Administration's Consumer Privacy Bill of Rights, released in 2012. It will be especially important to re-examine the traditional notice and consent framework that focuses on obtaining user permission prior to collecting data. While notice and consent remains fundamental in many contexts, it is now necessary to examine whether a greater focus on how data is used and reused would be a more productive basis for managing privacy rights in a big data environment. It may be that creating mechanisms for individuals to participate in the use and distribution of his or her information after it is collected is actually a better and more empowering way to allow people to access the benefits that derive from their information. Privacy protections must also evolve in a way that accommodates the social good that can come of big data use.

Advance the Consumer Privacy Bill of Rights

As President Obama made clear in February 2012, the Consumer Privacy Bill of Rights and the associated Blueprint for Consumer Privacy represent "a dynamic model of how to offer strong privacy protection and enable ongoing innovation in new information technologies." The Consumer Privacy Bill of Rights is based on the Fair Information Practice Principles. Some privacy experts believe nuanced articulations of these principles are flexible enough to address and support new and emerging uses of data, including big data. Others, especially technologists, are less sure, as it is undeniable that big data challenges several of the key assumptions that underpin current privacy frameworks, especially around collection and use. These big data developments warrant consideration in the context of how to viably ensure privacy protection and what practical limits exist to the practice of notice and consent.

> **Recommendation:** The Department of Commerce should promptly seek public comment on how the Consumer Privacy Bill of Rights could support the innovations of big data while at the same time responding to its risks, and how a responsible use framework, as articulated in Chapter 5, could be embraced within the framework established by the Consumer Privacy Bill of Rights.

> Following the comment process, the Department of Commerce should work on draft legislative text for consideration by stakeholders and for submission by the President to Congress.

Pass National Data Breach Legislation to Benefit Consumers and Businesses

As organizations store more information about individuals, Americans have a right to know if that information has been stolen or otherwise improperly exposed. A patchwork of 47 state laws currently governs when and how the loss of personally identifiable information must be reported.

> *Recommendation*: Congress should pass legislation that provides for a single national data breach standard along the lines of the Administration's May 2011 Cybersecurity legislative proposal. Such legislation should impose reasonable time periods for notification, minimize interference with law enforcement investigations, and potentially prioritize notification about large, damaging incidents over less significant incidents.

The Data Services Industry—Colloquially Known As "Data Brokers"—Should Bring Greater Transparency to the Sector

Consumers deserve more transparency about how their data is shared beyond the entities with which they do business directly, including "third-party" data collectors. This means ensuring that consumers are meaningfully aware of the spectrum of information collection and reuse as the number of firms that are involved in mediating their consumer experience or collecting information from them multiplies. The data services industry should follow the lead of the online advertising and credit industries and build a common website or online portal that lists companies, describes their data practices, and provides methods for consumers to better control how their information is collected and used or to opt-out of certain marketing uses.

Even As We Focus More on Data Use, Consumers Still Have a Valid Interest in "Do Not Track" Tools That Help Them Control When and How Their Data Is Collected

Strengthening these tools is especially important because there is now a growing array of technologies available for recording individual actions, behavior, and location data across a range of services and devices. Public surveys indicate a clear and overwhelming demand for these tools, and the government and private sector must continue working to evolve privacy-enhancing technologies in step with improved consumer services.

The Government Should Lead a Consultative Process to Assess How the Health Insurance Portability and Accountability Act and Other Relevant Federal Laws and Regulations Can Best Accommodate the Advances in Medical Science and Cost Reduction in Health Care Delivery Enabled By Big Data

Breakthroughs in predicting, detecting, and treating disease deserve the utmost public policy attention, but are unlikely to realize their full potential without substantial improvements in the medical data privacy regime that enables researchers to combine and analyze various kinds of lifestyle and health information. Any proposed reform must also consider bringing under regulatory and legal protection the vast quantities of personal health information circulated by organizations that are not covered entities governed by the Health Insurance Portability and Accountability Act.

The United States Should Lead International Conversations on Big Data That Reaffirms the Administration's Commitment to Interoperable Global Privacy Frameworks

The benefits of big data depend on the global free flow of information. The United States should engage international partners in a dialogue on the benefits and challenges of big data as they impact the legal frameworks and traditions of different nations.

Specifically, the Department of State and the Department of Commerce should actively engage with bilateral and intergovernmental partners, including the European Union, Asia Pacific Economic Cooperation (APEC), and Organization for Economic Cooperation and Development, and with other stakeholders, to take stock of how existing and proposed policy frameworks address big data.

The Administration should also work to strengthen the U.S.-European Union Safe Harbor Framework, encourage more countries and companies to join the APEC Cross Border Privacy Rules system, and promote collaboration on data flows between the United States, Europe and Asia through efforts to align Europe's system of Binding Corporate Rules and the APEC CBPR system.

Privacy Is a Worldwide Value That the United States Respects and Which Should Be Reflected in How It Handles Data Regarding All Persons

For this reason the United States should extend privacy protections to non-U.S. persons.

> **Recommendation:** The Office of Management and Budget should work with departments and agencies to apply the Privacy Act of 1974 to non-U.S. persons where practicable, or to establish alternative privacy policies that apply appropriate and meaningful protections to personal information regardless of a person's nationality.

2. Responsible Educational Innovation in the Digital Age

Big data offers significant opportunities to improve learning experiences for children and young adults. Big data intersects with education in two important ways. As students begin to share information with educational institutions, they expect that they are doing so in order to develop knowledge and skills, not to have their data used to build extensive profiles about their strengths and weaknesses that could be used to their disadvantage in later years. Educational institutions are also in a unique position to help prepare children, adolescents, and adults to grapple with the world of big data.

Ensure Data Protection While Promoting Innovation in Learning

Substantial breakthroughs stand to be made using big data to improve education as personalized learning on network-enabled devices becomes more common. Over the next five years, under the President's ConnectED initiative, American classrooms will receive a dramatic influx of technology—with substantial potential to enhance teaching and learning, particularly for disadvantaged communities. Internet-based education tools and software enable rapid iteration and innovation in educational technologies and businesses. These technologies are already being deployed with strong privacy and safety protections for students, inside and outside of the classroom. The Family Educational Rights and Privacy Act and Children's Online Privacy Protection Act provide a federal regulatory framework to protect the privacy of students—but FERPA was written before the Internet, and COPPA was written before smartphones, tablets, apps, the cloud, and big data. Students and their families need robust protection against current and emerging harms, but they also deserve access to the learning advancements enabled by technology that promise to empower all students to reach their full potential.

> **Recommendation:** The federal government should ensure that data collected in schools is used for educational purposes and continue to support investment and innovation that raises the level of performance across our schools. To promote this innovation, it should explore how to modernize the privacy regulatory framework under the Family Educational Rights and Privacy Act and Children's Online Privacy Protection Act and Children's Online Privacy Protection Act to ensure two complementary goals: 1) protecting students against their data being shared or used inappropriately, especially when that data is gathered in an educational context, and 2) ensuring that innovation in educational technology, including new approaches and business models, have ample opportunity to flourish.

Recognize Digital Literacy As an Important 21st Century Skill

In order to ensure students, citizens, and consumers of all ages have the ability to adequately protect themselves from data use and abuse, it is important that they develop fluency in understanding the ways in which data can be collected and shared, how algorithms are employed and for what purposes, and what tools and techniques they can use to protect themselves. Although such skills will never replace regulatory protections, increased digital literacy will better prepare individuals to live in a world saturated by data. Digital literacy—understanding how personal data is collected, shared, and used— should be recognized as an essential skill in K-12 education and be integrated into the standard curriculum.

3. Big Data and Discrimination

The technologies of automated decision-making are opaque and largely inaccessible to the average person. Yet they are assuming increasing importance and being used in contexts related to individuals' access to health, education, employment, credit, and goods and services. This combination of circumstances and technology raises difficult questions about how to ensure that discriminatory effects resulting from automated decision processes, whether intended or not, can be detected, measured, and redressed. We must begin a national conversation on big data, discrimination, and civil liberties.

The Federal Government Must Pay Attention to the Potential for Big Data Technologies to Facilitate Discrimination Inconsistent with the Country's Laws and Values

> **Recommendation:** The federal government's lead civil rights and consumer protection agencies, including the Department of Justice, the Federal Trade Commission, the Consumer Financial Protection Bureau, and the Equal Employment Opportunity Commission, should expand their technical expertise to be able to identify practices and outcomes facilitated by big data analytics that have a discriminatory impact on protected classes, and develop a plan for investigating and resolving violations of law in such cases. In assessing the potential concerns to address, the agencies may consider the classes of data, contexts of collection, and segments of the population that warrant particular attention, including for example genomic information or information about people with disabilities.

Consumers Have a Legitimate Expectation of Knowing Whether the Prices They Are Offered for Goods and Services Are Systematically Different Than the Prices Offered to Others

It is implausible for consumers to be presented with the full parameters of the data and algorithms shaping their online and offline experience. Nonetheless, some transparency is appropriate when a consumer's experience is being altered based on their personal information, particularly in situations where companies offer differential pricing to consumers in situations where they would not expect it—such as when comparing airline ticket prices on a web-based search engine or visiting the online storefront of a major retailer. The President's Council of Economic Advisers should assess the evolving practices of differential pricing both online and offline, assess the implications for efficient operations of markets, and consider whether new practices are needed to ensure fairness for the consumer.

Data Analytics Can Be Used to Shore up Civil Liberties

The same big data technologies that enable discrimination can also help groups enforce their rights. Applying correlative and data mining capabilities can identify and empirically confirm instances of discrimination and characterize the harms they caused. The federal government's civil rights offices, together with the civil rights community, should employ the new and powerful tools of big data to ensure that our most vulnerable communities are treated fairly.

To build public awareness, the federal government's consumer protection and technology agencies should convene public workshops and issue reports over the next year on the potential for discriminatory practices in light of these new technologies; differential pricing practices; and the use of proxy scoring to replicate regulated scoring practices in credit, employment, education, housing, and health care.

4. Law Enforcement and Security

Big data, lawfully applied, can make our communities safer, make our nation's infrastructure more resilient, and strengthen our national security. It is crucial that the national security, homeland security, law enforcement, and intelligence communities continue to vigorously experiment with and apply lawful big data technology while adhering to full accountability, oversight, and relevant privacy requirements.

The Electronic Communications Privacy Act Should Be Reformed

> **Recommendation:** Congress should amend ECPA to ensure the standard of protection for online, digital content is consistent with that afforded in the physical world—including by removing archaic distinctions between email left unread or over a certain age.

The Use of Predictive Analytics by Law Enforcement Should Continue to Be Subjected to Careful Policy Review

It is essential that big data analysis conducted by law enforcement outside the context of predicated criminal investigations be deployed with appropriate protections for individual privacy and civil liberties. The presumption of innocence is the bedrock of the American criminal justice system. To prevent chilling effects to Constitutional rights of free speech and association, the public must be aware of the existence, operation, and efficacy of such programs.

Federal Agencies with Expertise in Privacy and Data Practices Should Provide Technical Assistance to State, Local, and Other Federal Law Enforcement Agencies Seeking to Deploy Big Data Techniques

Law enforcement agencies should continue to examine how federal grants involving big data surveillance technologies can foster their responsible use, as well as the potential utility of establishing a national registry of big data pilots in state and local law enforcement in order to track, identify, and promote best practices. Federal government agencies with technology leaders and experts should also report progress in developing privacy-protective technologies over the next year to help advance the development of technical skills for the advancement of the federal privacy community.

Government Use of Lawfully-Acquired Commercial Data Should Be Evaluated to Ensure Consistency with Our Values

Recognizing the longstanding practice of basic commercial records searches against criminal suspects, the federal government should undertake a review of uses of commercially available data on U.S. citizens, focusing on the use of services that employ big data techniques and ensuring that they incorporate appropriate oversight and protections for privacy and civil liberties.

Federal Agencies Should Implement Best Practices for Institutional Protocols and Mechanisms That Can Help Ensure the Controlled Use and Secure Storage of Data

The Department of Homeland Security, the intelligence community, and the Department of Defense are among the leaders in developing privacy-protective technologies and policies for handling personal data. Other public sector agencies should evaluate whether any of these practices—particularly data tagging to enforce usage limitations, controlled access policies, and immutable auditing—could be integrated into their databases and data practices to provide built-in protections for privacy, civil rights, and civil liberties.

Use Big Data Analysis and Information Sharing to Strengthen Cybersecurity

Protecting the networks that drive our economy, sustain public safety, and protect our national security has become a critical homeland security mission. The federal government's collaboration with private sector partners to use big data in programs, pilots, and research for both cybersecurity and protecting critical infrastructure can help strengthen our resilience and cyber defenses, especially as more cyber threat data is shared. The Administration continues to support legislation that protects privacy while providing targeted liability protection for companies sharing certain threat information and appropriately defending their networks on that basis. At the same time, the Administration will continue to use executive action to increase incentives for and reduce barriers to the kind of information sharing and analytics that will help the public and private sector prevent and respond to cyber threats.

5. Data As a Public Resource

Government data is a national resource, and should be made broadly available to the public wherever possible, to advance government efficiency, ensure government accountability, and generate economic prosperity and social good—while continuing to protect personal privacy, business confidentiality, and national security. This means finding new opportunities for the government to release large data sets and ensuring all agencies make maximum use of Data.gov, a repository of federal data tools and resources. Big data can help improve the provision of public services, provide new insights to inform policymaking, and increase the efficient use of taxpayer dollars at every level of government.

Government Data Should Be Accurate and Securely Stored, and to the Maximum Extent Possible, Open and Accessible

Government data—particularly statistical and census data—distinguishes itself by providing a high level of accuracy, reliability, and confidentiality. Similarly, the "My Data" initiatives that currently allow Americans easy, secure access to their own digital data in useful formats constitutes a model for personal data accessibility that should be replicated as widely as possible across the government.

All Departments and Agencies Should, in Close Coordination with Their Senior Privacy and Civil Liberties Officials, Examine How They Might Best Harness Big Data to Help Carry out Their Missions

Departments and agencies that have not historically made wide use of advanced data analytics should make the most out of what the big data revolution means for them and the citizens they serve. They should experiment with pilot projects, develop in-house talent, and potentially expand research and development. From the earliest stages, agencies should build these projects in consultation with their privacy and civil liberties officers.

In particular, big data analytics present an important opportunity to increase value and performance for the American people in the delivery of government services.

Big data also holds enormous power to detect and address waste, fraud and abuse, thereby saving taxpayer money and improving public trust.

Big data can further help identify high performers across government whose practices can be replicated by similar agencies and programs and may deliver new insights into effective public-sector management.

We Should Dramatically Increase Investment for Research and Development in Privacy-Enhancing Technologies, Encouraging Cross-Cutting Research That Involves Not Only Computer Science and Mathematics, but Also Social Science, Communications and Legal Disciplines

The Administration should lead an effort to identify areas where big data analytics can provide the greatest impact for improving the lives of Americans and encourage data scientists to develop social, ethical, and policy knowledge. To this end, the Office of Science and Technology Policy, in partnership with experts across the agencies, should work to define areas that promise significant public gains—for example, in urban informatics—and assess how to provide appropriate attention and resources.

Promising areas for basic research include data provenance, de-identification and encryption, but we also encourage focusing on lab-to-market tools that can be rapidly deployed to consumers. Because we will need a growing cadre of data and social scientists who are able to encode critical policy values into technical infrastructure, we support investment in fields such as Science and Technology Studies which emphasize teaching scientific knowledge and technology in its social and ethical context, and the teaching of module courses to data scientists and engineers to familiarize them with the broader societal implications of their work.

APPENDIX

A. Methodology

This 90-day study was announced by President Obama in his January 17, 2014 remarks on the review of signals intelligence. He charged his Counselor John Podesta to "look how the challenges inherent in big data are being confronted by both the public and private sectors; whether we can forge international norms on how to manage this data; and how we can continue to promote the free flow of information in ways that are consistent with both privacy and security." Podesta led a working group of senior Administration officials including Secretary of Commerce Penny Pritzker, Secretary of Energy Ernie Moniz, Director of the Office of Science and Technology Policy John Holdren, and Director of the National Economic Council Jeffrey Zients. Nicole Wong, R. David Edelman, Christopher Kirchhoff, and Kristina Costa were the principal staff authors supporting this report. To inform its deliberations, the working group initiated a broad public dialogue on the implications of technological advancements in big data.

During the course of this study, the working group met with hundreds of stakeholders from industry, academia, civil society, and the federal government through briefings at the White House. These briefings provided a chance for dialogue with key stakeholders, including privacy and civil liberties advocates; scientific and statistical agencies; international data protection authorities; the intelligence community; law enforcement officials; leading academics who study social and technical aspects of privacy and the Internet; and practitioners and executives from the health care, financial, and information services industries. A full list of briefings and participants is included in Section B of the appendix.

To further engage the public, the White House Office of Science and Technology Policy sponsored conferences at the Massachusetts Institute of Technology, New York University, and the University of California, Berkeley. Senior Administration officials, including Counselor Podesta and Secretary Pritzker, participated in these conferences, along with policy experts, academics, and representatives from business and the nonprofit community. Details of these conferences and a list of presentations is included in Section C of the appendix.

The working group also published a Federal Register notice to gather written input, and used the whitehouse.gov platform to solicit comments from the general public online. Details of these efforts are included in Sections E and F of the appendix.

B. Stakeholder Meetings

Acxiom
Adobe
Allstate
Ally Financial
Amazon
American Association of Advertising Agencies
American Association of Universities
American Civil Liberties Union
Apple
AppNexus
Archimedes Incorporated
Asian Americans Advancing Justice
Association of National Advertisers
athenahealth
Bank of America
BlueKai
Bureau of Consumer Protection
Canadian Interim Privacy Commissioner
Capital One
Carnegie Mellon University
Cato Institute
Census Bureau
Center for Democracy & Technology
Center for Digital Democracy
Center for National Security Studies
Central Intelligence Agency
ColorOfChange
Computer Science and Artificial Intelligence Laboratory, MIT
comScore
Corelogic
Cornell University
Council of Better Business Bureaus
Datalogix

Department of Commerce, General Counsel
Department of Homeland Security
Digital Advertising Alliance
Direct Marketing Association
Discover
Drug Enforcement Administration
Duke University School of Law
Dutch Data Protection Authority
Economics and Statistics Administration Electronic Frontier Foundation
Electronic Privacy Information Center
Epsilon
European Union Data Protection Supervisor
European Commission: Directorate-General for Justice (Data Protection Division)
Evidera
Experian
Explorys
Facebook
Federal Bureau of Investigation
Federal Telecommunications Commission, Bureau of Consumer Protection
Financial Services Roundtable
Free Press
French National Commission on Informatics and Liberty
Future of Privacy
George Washington University
Georgetown University Law Center
GNS Health care
Google
GroupM
Harvard University
Humedica
IBM Health Care
IMS Health
Infogroup
Interactive Advertising Bureau
International Association of Privacy Professionals
Jenner & Block LLP
Lawrence Berkeley National Laboratory

Lawrence Livermore National Laboratory
LexisNexis
LinkedIn
Massachusetts Institute of Technology
Massachusetts Institute of Technology Media Lab
MasterCard
Mexican Data Privacy Commissioner
Microsoft
National Association for the Advancement of Colored People
National Economic Council
National Hispanic Media Coalition
National Oceanic and Atmospheric Administration
National Organization for Women
National Security Agency
National Telecommunications and Information Administration
National Urban League Policy Institute
NaviMed Capital
Network Advertising Initiative
Neustar
Office of Chairwoman Edith Ramirez
Office of Science and Technology Policy
Office of the Director of National Intelligence
Office of the National Coordinator for Health Information Technology
Ogilvy
Open Society Foundations
Open Technology Institute
Optum Labs
PatientsLikeMe
Princeton University
Privacy Analytics
Public Knowledge
Quantcast
Robinson & Yu LLC
SalesForce
The Brookings Institution
The Constitution Project
The Leadership Conference on Civil and Human Rights
UK Information Commissioner
University of Maryland

University of Virginia
Visa
Yahoo!
Zillow

C. Academic Symposia

Big Data and Privacy Workshop
Advancing the State of the Art in Technology and Practice *Massachusetts Institute of Technology (MIT)*
Cambridge, Massachusetts
March 3, 2014

    Welcome: L. Rafael Reif, President of MIT
    Keynote: John Podesta, Counselor to the President
    Keynote: Penny Pritzker, Secretary of Commerce
    State of the Art of Privacy Protection: Cynthia Dwork, Microsoft

    Panel Session 1: Big Data Opportunities and Challenges
    Panel Chair: Daniela Rus, MIT
    Mike Stonebraker, MIT
    John Guttag, MIT
    Manolis Kellis, MIT
    Sam Madden, MIT
    Anant Agarwal, edX

    Panel Session 2: Privacy Enhancing Technologies
    Panel Chair: Shafi Goldwasser
    Nickolai Zeldovich, MIT
    Vinod Vaikuntanathan, Assistant Professor, MIT
    Salil Vadhan, Harvard University
    Daniel Weitzner, MIT

    Panel Session 3: Roundtable Discussion of Large-Scale Analytics Case Study
    Panel Moderator: Daniel Weitzner
    Chris Calabrese, American Civil Liberties Union
    John DeLong, National Security Agency

Mark Gorenberg, Zetta Venture Partners
David Hoffman, Intel
Karen Kornbluh, Nielsen
Andy Palmer, KOA Lab
James Powell, Thomson Reuters
Latanya Sweeney, Harvard University
Vinod Vaikuntanathan, MIT

Concluding Statements: Maria Zuber, MIT

The Social, Cultural, & Ethical Dimensions of 'Big Data'
The Data & Society Research Institute & New York University (NYU)
New York, New York
March 17, 2014

Introduction: danah boyd, Data & Society
Fireside Chat: John Podesta, Counselor to the President
Keynote: Penny Pritzker, Secretary of Commerce
State of the Art of Privacy Protection: Cynthia Dwork, Microsoft
Discussion Breakouts
    Tim Hwang: On Cognitive Security
    Nick Grossman: Regulation 2.0
    Nuala O'Connor: The Digital Self & Technology in Daily Life
    Alex Howard: Data Journalism in the Second Machine Age
    Mark Latonero: Big Data and Human Trafficking
    Corrine Yu: Civil Rights Principles for the Era of Big Data
    Natasha Schüll: Tracking for Profit; Tracking for Protection
    Kevin Bankston: The Biggest Data of All
    Alessandro Acquisti: The Economics of Privacy (and Big Data)
    Latanya Sweeney: Transparency Builds Trust
    Deborah Estrin: You + Your Data
    Clay Shirky: Analog Thumbs on Digital Scales Open Discussion
    Moderators: danah boyd and Nicole Wong
Workshops
    Data Supply Chains
    Inferences and Connections
    Predicting Human Behavior
    Algorithmic Accountability
    Interpretation Gone Wrong

Inequalities and Asymmetries
Public Plenary
    Welcome: danah boyd, Data & Society
    Video Address: John Podesta, Counselor to the President
    Keynote: Nicole Wong, Deputy Chief Technology Officer of the US
    Plenary Panel Statements
        Kate Crawford, Microsoft Research and MIT
        Anil Dash, Think Up and Activate (moderator)
        Steven Hodas, NYC Department of Education
        Alondra Nelson, Columbia University
        Shamina Singh, MasterCard Center for Inclusive Growth

Big Data: Values and Governance
University of California, Berkeley (UC Berkeley)
Berkeley, California
April 1, 2014

Welcome: Dean AnnaLee Saxenian, UC Berkeley School of Information
Welcome: Nicole Wong, Deputy Chief Technology Officer, OSTP

Panel Session 1: Values at stake, Values in tension: Privacy and Beyond
    Moderator: Deirdre Mulligan, UC Berkeley School of Information
    Amalia Deloney, Center for Media Justice
    Nicole Ozer, Northern California ACLU
    Fred Cate, University of Indiana
    Kenneth A. Bamberger, UC Berkeley School of Law
Panel Session 2: New Opportunities and Challenges in Health and Education
    Moderator: Paul Ohm, University of Colorado Law School
    Barbara Koenig, University of California, San Francisco
    Deven McGraw, Center for Democracy & Technology
    Scott Young, Kaiser Permanente
    Zachary Pardos, UC Berkeley School of Information
Panel Session 3: Algorithms: Transparency, Accountability, Values and Discretion
    Moderator: Omer Tene, International Association of Privacy Professionals
    Ari Gesher, Palantir
    Lee Tien, Electronic Frontier Foundation

Seeta Gangadharan, New America Foundation
Thejo Kote, Automatic
James Rule, UC Berkeley
Governance Roundtable
Moderator: David Vladeck, Georgetown University Law School
Julie Brill, Federal Trade Commission
Erika Rottenberg, LinkedIn
Cameron Kerry, MIT Media Lab
Cynthia Dwork, Microsoft Research
Mitchell Stevens, Stanford University
Rainer Stentzel, German Federal Ministry of the Interior

Concluding Keynote: John Podesta, Counselor to the President

D. PCAST Report

To take measure of the shifting technological landscape, the President charged his Council of Advisors on Science & Technology (PCAST) to conduct a parallel study to assess the technological dimensions of the intersection of big data and privacy. PCAST's statement of work reads, in part:

> "PCAST will study the technological aspects of the intersection of big data with individual privacy, in relation to both the current state and possible future states of the relevant technological capabilities and associated privacy concerns.
> Relevant big data include data and metadata collected, or potentially collectable, from or about individuals by entities that include the government, the private sector, and other individuals. It includes both proprietary and open data, and also data about individuals collected incidentally or accidentally in the course of other activities (e.g., environmental monitoring or the "Internet of things")."

The PCAST assessment was conducted simultaneously with the 90-study on big data.

PCAST shared their preliminary conclusions with the working group in order to inform its deliberations.

E. Public Request for Information

As part of the effort to make this review as inclusive as possible, the White House Office of Science and Technology Policy (OSTP) released a Request for Information (RFI) seeking public comment on the ways in which big data may impact privacy, the economy, and public policy. The RFI was published on March 4, 2014, and 76 comments were submitted through April 4, 2014. The comments came from nonprofits, corporations, universities, and individual citizens. The full list of respondents is included below, and the full text of all responses is publicly available at whitehouse.gov/bigdata.

The RFI Posed Five Questions to Respondents:
(1) What are the public policy implications of the collection, storage, analysis, and use of big data? For example, do the current U.S. policy framework and privacy proposals for protecting consumer privacy and government use of data adequately address issues raised by big data analytics?
(2) What types of uses of big data could measurably improve outcomes or productivity with further government action, funding, or research? What types of uses of big data raise the most public policy concerns? Are there specific sectors or types of uses that should receive more government and/or public attention?
(3) What technological trends or key technologies will affect the collection, storage, analysis and use of big data? Are there particularly promising technologies or new practices for safeguarding privacy while enabling effective uses of big data?
(4) How should the policy frameworks or regulations for handling big data differ between the government and the private sector? Please be specific as to the type of entity and type of use (e.g., law enforcement, government services, commercial, academic research, etc.).
(5) What issues are raised by the use of big data across jurisdictions, such as the adequacy of current international laws, regulations, or norms?

The RFI can be found at:
http://www.gpo.gov/fdsys/pkg/FR-2014-03-04/pdf/2014-04660.pdf.

Respondents
Access
American Civil Liberties Union

Ad Self-Regulatory Council, Council of Better Business Bureaus
Annie Shebanow
The Architecture for a Digital World and Advanced Micro Devices
Association for Computing Machinery
Association of National Advertisers
Brennan Center for Justice
BSA | The Software Alliance
Center for Democracy and Technology
Center for Data Innovation
Center for Digital Democracy
Center for National Security Studies
Cloud Security Alliance
Coalition for Privacy and Free Trade
Common Sense Media
Computer and Communications Industry Association
Computing Community Consortium
Constellation Research
Consumer Action
Consumer Federation of America
Consumer Watchdog
Dell
Direct Marketing Association
Dr. Tyrone W A Grandison
Dr. A. R. Wagner
Durrell Kapan
Electronic Frontier Foundation
Electronic Transactions Association
Entity
Federation of American Societies for Experimental Biology
Financial Services Roundtable
Food Marketing Groups
Frank Pasquale, UMD Law
Fred Cate, Microsoft, Oxford Internet Institute
Future of Privacy Forum
Georgetown University
Health care Leadership Council
IMS Health
Information Technology Industry Council
Interactive Advertising Bureau

Intrical
IT Law Group
Jackamo
James Cooper, George Mason Law
Jason Kint
Jonathan Sander, STEALTHbits
Kaliya Identity Woman
Leadership Conferences on Civil and Human Rights & Education
Making Change at Walmart
Marketing Research Association
Mary Culnan, Bentley University & Future of Privacy Forum
McKenna Long & Aldridge LLP
mediajustice.org
Microsoft
Massachusetts Institute of Technology
MITRE Corporation
Mozilla
New York University Center for Urban Science & Progress
Online Trust Alliance
Pacific Northwest National Laboratory
Peter Muhlberger
Privacy Coalition
Reed Elsevier
Sidley Austin LLP
Software & Information Industry Association
TechAmerica
TechFreedom
Technology Policy Institute
The Internet Association
U.S. Chamber of Commerce
U.S. Leadership for the Revision of the 1967 Space Treaty
U.S. PIRG VIPR Systems
World Privacy Forum

F. White House Big Data Survey

Additional public input about big data and privacy issues was solicited via a short web form posted on WhiteHouse.gov and promoted via email and social media. During the four weeks the survey was open for public input, 24,092 people submitted responses. It is important to note, however, that this process was a means of gathering public input and should not be considered a statistically representative survey of attitudes about data privacy. The White House did not include submission fields for name or contact information on the survey form.

Respondents expressed a great deal of concern about big data practices. They communicated particularly strong feelings around ensuring that data practices have proper transparency and oversight—more than 80 percent of respondents were very concerned with each of these areas—but even in the area of least concern (collection of location data), 61 percent indicated that they were "very much concerned" about this practice. By contrast, considerably more nuance was evident in respondents' views towards particular entities. Although majorities claimed to trust Intelligence and Law Enforcement Agencies "not at all," their views towards other government agencies at both federal and local levels were far less negative. Furthermore, majorities were generally trusting of how professional practices, like law and medical offices, and academia use and handle big data.

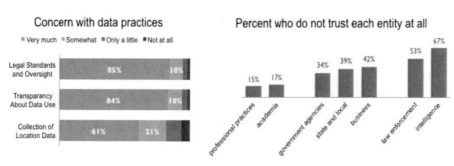

Taken together, the findings from this survey indicate that respondents were most wary of how intelligence and law enforcement agencies are collecting and using data about them, particularly when they have little insight into these practices. This suggests that the Administration should work to increase the transparency about intelligence practices where possible, reassure the public that collected data is stored as securely as possible, and strengthen applicable legal structures and oversight.

For more information about the survey, visit: WhiteHouse.gov/BigData.

End Notes

[1] Scott Crosier, *John Snow: The London Cholera Epidemic of 1854,* Center for Spatially Integrated Social Science, University of California, Santa Barbara, 2007, http://www.csiss.org/classics/content/8.

[2] Simon Head, *The New Ruthless Economy: Work and Power in the Digital Age,* (Oxford University Press, 2005).

[3] Frederick Taylor, *The Principles of Scientific Management* (Harper & Brothers, 1911), p. 7, http://www.eldritchpress.org/fwt/ti.html.

[4] John Gantz and David Reinsel, *Extracting Value from Chaos,* IDC, 2011, http://www.emc.com/collateral/analyst-reports/idc-extracting-value-from-chaos-ar.pdf.

[5] Mary Meeker and Liang Yu, *Internet Trends,* Kleiner Perkins Caulfield Byers, 2013, http://www.slideshare.net/kleinerperkins/kpcb-internet-trends-2013.

[6] "2016: The Year of the Zettabyte," Daily Infographic, March 23, 2013, http://dailyinfographic.com/2016-theyear-of-the-zettabyte-infographic.

[7] Liran Einav and Jonathan Levin, "The Data Revolution and Economic Analysis," Working Paper, No. 19035, *National Bureau of Economic Research,* 2013, http://www.nber.org/papers/w19035; Viktor Mayer-Schonberger and Kenneth Cukier, *Big Data: A Revolution That Will Transform How We Live, Work, and Think,* (Houghton Mifflin Harcourt, 2013).

[8] National Science Foundation, Solicitation 12-499: *Core Techniques and Technologies for Advancing Big Data Science & Engineering (BIGDATA),* 2012, http://www.nsf.gov/pubs/2012/nsf12499/nsf12499.pdf.

[9] Harvard Professor of Science & Technology Studies Sheila Jasanoff argues that framing the policy implications of big data is difficult precisely because it manifests in multiple contexts that each call up different operative concerns, including big data as property (who owns it); big data as common pool resources (who manages it and on what principles); and big data as identity (it is us ourselves, and thus its management raises constitutional questions about rights).

[10] President's Council of Advisors on Science & Technology, *Big Data and Privacy: A Technological Perspective,* The White House, May 1, 2014.

[11] The distinction between data that is "born analog" and data that is "born digital" is explored at length in the PCAST report, *Big Data and Privacy,* p 18-22.

[12] See, e.g., Kapow Software, *Intelligence by Variety Where to Find and Access Big Data,* http://www.kapowsoftware.com/resources/infographics/intelligence-by-variety-where-to-find-and-access-bigdata.php; James Manyika, Michael Chui, Brad Brown, Jacques Bughin, Richard Dobbs, Charles Roxburgh, and Angela Hung Byers, *Big Data: The Next Frontier for Innovation, Competition, and Productivity,* McKinsey Global Institute, 2011, http://www.mckinsey.com/insights/business_technology/big_data_the_next_frontier_for_in novation.

[13] Salesforce.com, "Collaboration helps GE Aviation bring its best inventions to life," http://www.salesforce.com/customers/stories/ge.jsp; Armand Gatdula, "Fleet Tracking Devices will be Installed in 22,000 UPS Trucks to Cut Costs and Improve Driver Efficiency in 2010," FieldLogix.com blog, July 20, 2010, http://www.fieldtechnologies.com/gps-tracking-systems-installed-in-ups-trucks-driver-efficiency.

[14] The Patient Protection and Affordable Care Act provides additional resources for fraud prevention. Centers for Medicare and Medicaid Services, "Fraud Prevention Toolkit," http://www.cms.gov/Outreach-andEducation/Outreach/Partnerships/FraudPrevention Toolkit.html.

[15] IBM, "Smarter Healthcare in Canada: Redefining Value and Success," July 2012, http://www.ibm.com/smarterplanet/global/files/ca__en_us__health care__ca_brochure.pdf.

[16] Manolis Kellis, "Importance of Access to Large Populations," *Big Data Privacy Workshop: Advancing the State of the Art in Technology and Practice*, Cambridge, MA, March 3, 2014, http://web.mit.edu/bigdatapriv/ppt/ManolisKellis_PrivacyBigData_CSAIL-WH.pptx.

[17] Latanya Sweeney, "Discrimination in Online Ad Delivery," 2013, http://dataprivacylab.org/projects/onlineads/1071-1.pdf.

[18] Cynthia Dwork and Deirdre Mulligan, "It's Not Privacy, and It's Not Fair," 66 *Stan. L. Rev. Online* 35 (2013).

[19] See PCAST report, *Big Data and Privacy*; Harvard Law Petrie-Flom Center, *Online Symposium on the Law, Ethics & Science of Re-identification Demonstrations*, http://blogs.law.harvard.edu/billofhealth/2013/05/13/online-symposium-on-the-law-ethics-science-of-re-identification-demonstrations/.

[20] President Barack Obama, *International Strategy for Cyberspace*, The White House, May 2011, http://www.whitehouse.gov/the-press-office/2014/02/12/launch-cybersecurity-framework.

[21] President Barack Obama, Remarks on the Administration's Review of Signals Intelligence, January 17, 2014, http://www.whitehouse.gov/the-press-office/2014/01/17/remarks-president-review-signals-intelligence.

[22] Melvin Kranzberg, "Technology and History: Kranzberg's Laws," 27.3 *Technology and Culture*, (1986) p.544-560.

[23] See, e.g., *City of Ontario v. Quon*, 560 U.S. 746, 755-56 (2010) ("The [Fourth] Amendment guarantees the privacy, dignity, and security of persons against certain arbitrary and invasive acts by officers of the Government."); *Kyllo v. United States*, 533 U.S. 27, 31 (2001) ("'At the very core' of the Fourth Amendment 'stands the right of a man to retreat into his own home and there be free from unreasonable governmental intrusion.'"); *Olmstead v. United States*, 277 U.S. 438, 478 (1928) (Brandeis, J., dissenting) ("They [the Framers] sought to protect Americans in their beliefs, their thoughts, their emotions and their sensations. They conferred, as against the Government, the right to be let alone—the most comprehensive of rights and the right most valued by civilized men.").

[24] For example, e.g. the 1790 Census counted white men "over 16" and "under 16" separately to determine military eligibility. United States Census Bureau, "History," https://www.census.gov/history/www/through_the_decades/index_of_questions/1790_1.html; Margo Anderson, *The American Census: A Social History*, (Yale University Press, 1988).

[25] President Barack Obama, *Consumer Data Privacy In A Networked World: A Framework For Protecting Privacy And Promoting Innovation In The Global Digital Economy*, The White House, February 2012, http://www.whitehouse.gov/sites/default/files/privacy-final.pdf.

[26] National Institute of Standards & Technology, *Framework for Improving Critical Infrastructure Cybersecurity*, February 12, 2014, http://www.nist.gov/cyberframework/upload/cybersecurity-framework-021214-final.pdf.

[27] President Barack Obama, Making Open and Machine Readable the New Default for Government Information, Executive Order 13642, May 2013, http://www.whitehouse.gov/the-pressoffice/2013/05/09/executive-order-making-open-and-machine-readable-new-default-government.

[28] Office of Management and Budget, Guidance for Providing and Using Administrative Data for Statistical Purposes, (OMB M-144-06), February 14, 2014, http://www.whitehouse.gov/sites/default/files/omb/memoranda/2014/m-14-06.pdf.

[29] These events have helped federal agencies showcase government data resources being made freely available; collaborate with innovators about how open government data can be used to fuel new products, services, and companies; launch new challenges and incentive prizes designed to spur innovative use of data; and highlight how new uses of open government data are making a tangible impact in American lives and advancing the national interest.

[30] Specifically, the Open Data Policy (OMB M-13-13) requires agencies to collect or create information in a way that supports downstream information processing and dissemination; to maintain internal and external data asset inventories; and to clarify information

management responsibilities. Agencies must also use machine-readable and open formats, data standards, and common core and extensible metadata.

[31] Aneesh Chopra, "Green Button: Providing Consumers with Access to Their Energy Data," *Office of Science and Technology Policy Blog*, January 2012, http://www.whitehouse.gov/blog/2012/01/18/green-button-providing-consumers-access-their-energy-data.

[32] In November 2013, the White House organized a "Data to Knowledge to Action" event that featured dozens of announcements of new public, private, academic and non-profit initiatives. From transforming how research universities prepare students to become data scientists to allowing more citizens and entrepreneurs to access and analyze the huge amounts of space-based data that NASA collects about the Earth, the commitments promise to spur tremendous progress. The Administration is also working to increase the number of data scientists who are actively engaged in solving hard problems in education, health care, sustainability, informed decision-making, and non-profit effectiveness.

[33] Samuel Warren and Louis Brandeis, "The Right to Privacy," 4 *Harvard Law Review* 193, 195 (1890).

[34] See William Prosser, "Privacy," 48 *California Law Review* 383 (1960).

[35] Wayne Lafave, "Search and Seizure: A Treatise On The Fourth Amendment," §§ 1.1–1.2 (West Publishing, 5th ed. 2011).

[36] *Olmstead v. United States*, 277 U.S. 438 (1928).

[37] Ibid at 478.

[38] *Katz v. United States*, 389 U.S. 347, 361 (1967) (Harlan, J., concurring); see also LaFave, supra note 35 § 2.1(b) ("[L]ower courts attempting to interpret and apply Katz quickly came to rely upon the Harlan elaboration, as ultimately did a majority of the Supreme Court.").

[39] Restatement (First) Torts § 867 (1939).

[40] Prosser, supra note 34 at 389 (1960).

[41] Ibid. See also Restatement (Second) Torts § 652A (1977) (Prosser's privacy torts incorporated into the Restatement).

[42] Ibid.

[43] See, e.g., K.A. Taipale, "Data Mining and Domestic Security: Connecting the Dots to Make Sense of Data," V *The Columbia Science and Technology Review*, (2003), http://papers.ssrn. com/sol3/papers.cfm?abstract_id=546782.

[44] Pub. L. 93-579 (codified at 5 U.S.C. § 552a).

[45] Organization for Economic Cooperation and Development, *Thirty Years After The OECD Privacy Guidelines*, 2011, p. 17, http://www.oecd.org/sti/ieconomy/49710223.pdf.

[46] Ibid at 27.

[47] The APEC Privacy Principles are associated with the 2004 APEC Privacy Framework and APEC Cross Border Privacy Rules system approved in 2011. See Asia-Pacific Economic Cooperation, "APEC Privacy Principles," 2005, p. 3, http://www.apec.org/Groups/Committee-on-Trade-andInvestment/~/media/Files/Groups/ECSG/05_ecsg_privacyframe wk.ashx; *Consumer Data Privacy In A Networked World*, p 49-52; export.gov/safeharbor for information on the U.S.-EU and U.S.-Swiss Safe Harbor Frameworks. These enforceable self-certification programs are administered by the U.S. Department of Commerce and were developed in consultation with the European Commission and the Federal Data Protection and Information Commissioner of Switzerland, respectively, to provide a streamlined means for U.S. organizations to comply with EU and Swiss data protection laws.

[48] California, for example, has a right to privacy in the state Constitution. Cal. Const. art. 1 § 1.

[49] See U.S. Department of Health and Human Services, Health Information Privacy, "Summary of the HIPAA Privacy Rule," http://www.hhs.gov/ocr/privacy/hipaa/understanding/summary/index.html

[50] This principle ensures that covered entities make reasonable efforts to use, disclose, and request only the minimum amount of protected health information needed to accomplish the intended purpose of the use, disclosure, or request. See U.S. Department of Health &

Human Services, Health Information Privacy, "Minimum Necessary Requirement," http://www.hhs.gov/ocr/privacy/hipaa/understanding/coveredentities/minimumnecessary.html.

[51] They include: The Fair Credit Reporting Act of 1970, the Family Educational Rights and Privacy Act of 1974, the Electronic Communications Privacy Act of 1986, the Computer Fraud and Abuse Act of 1986, the Cable Communications Policy Act of 1984, the Video Privacy Protection Act of 1998, and the Genetic Information Nondiscrimination Act of 2008.

[52] See *Consumer Data Privacy In A Networked World*, p 25.

[53] Organization for Economic Cooperation and Development, "OECD Work on Privacy," http://www.oecd.org/sti/ieconomy/privacy.htm.

[54] European Commission, "Commission Proposes a Comprehensive Reform of the Data Protection Rules," January 25, 2012, http://ec.europa.eu/justice/newsroom/data-protection/news/120125_en.htm.

[55] See Joined Cases C-293/12 and C-594/12, Digital Rights Ireland Ltd. v. Minister for Communications, Marine and Natural Resources, et al. (Apr. 8, 2014) in which the European Court of Justice invalidated the data retention requirements applied to electronic communications on the basis that the scope of the requirements interfered in a "particularly serious manner with the fundamental rights to respect for private life and to the protection of personal data."

[56] European Commission, Article 29 Data Protection Working Party, Press Release: "Promoting Cooperation on Data Transfer Systems Between Europe and the Asia-Pacific," March 26, 2013, http://ec.europa.eu/justice/data-protection/article-29/press-material/press-release/art29_press_material/20130326_pr_apec_en.pdf.

[57] Article 29 Data Protection Working Party, Opinion 02/2014 on a referential for requirements for Binding Corporate Rules, February 27, 2014, http://ec.europa.eu/justice/data-protection/article-29/documentation/opinion-recommendation/files/2014/wp212_en.pdf.

[58] Bradley Malin and Latanya Sweeney, "How (not) to protect genomic data privacy in a distributed network: using trail re-identification to evaluate and design anonymity protection systems," *Journal of Biomedical Informatics* (2004), http://www.j-biomed-inform.com/article/S1532-0464(04)00053-X.

[59] Latanya Sweeney, a Professor of Government and Technology in Residence at Harvard University, has studied information flows in the health care industry. A graphical map of data flows that depicts information flows outside entities regulated by HIPAA can be found at www.thedatamap.org.

[60] President's Council of Advisors on Science & Technology, *Realizing the Full Potential Of Health Information Technology to Improve Health Care for Americans: The Path Forward*, The White House, December 2010, http://www.whitehouse.gov/sites/default/files/microsites/ostp/pcast-health-it-report.pdf.

[61] President's Council of Advisors on Science & Technology, *Harnessing Technology for Higher Education*, The White House, December 2013, http://www.whitehouse.gov/sites/default/files/microsites/ostp/PCAST/pcast_edit_dec-2013.pdf.

[62] Department of Education, *Enhancing Teaching and Learning Through Educational Data Mining and Learning Analytics: An Issue Brief*, October 2012, http://www.ed.gov/edblogs/technology/files/2012/03/edmla-brief.pdf. For information about the National Education technology plan, see www.tech.ed.gov/netp.

[63] danah boyd, *It's Complicated: The Social Lives of Networked Teens*, (Yale University Press, 2014), www.danah.org/books/ItsComplicated.pdf.

[64] Department of Education, *Protecting Student Privacy While Using Online Educational Services: Requirement and Best Practices*, February 2014, http://ptac.ed.gov/sites/default/files/Student%20Privacy%20and%20Online%20Educational%20Services%20%28February%202014%29.pdf.

[65] For example, California recently passed a law prohibiting online services from gathering information about a minor's activities for marketing purposes, or from displaying certain

online advertising to minors. The law further requires online services to delete information that the minor posted on the website or service, a right for which the statute has now been dubbed "the Eraser Law."

[66] The Department of Education is exploring data innovation and use in a wide variety of contexts, including making more educational data available through application programming interfaces. See David Soo, "How can the Department of Education Increase Innovation, Transparency and Access to Data?," *Department of Education Blog*, http://www.ed.gov/blog/2014/04/how-can-the-department-of-education-increase-innovation-transparency-and-access-to-data/.

[67] Department of Education, Technology in Education: Privacy and Progress, Remarks of U.S. Secretary of Education Arne Duncan at the Common Sense Media Privacy Zone Conference, February 24, 2014, https://www.ed.gov/news/speeches/technology-education-privacy-and-progress.

[68] Department of Homeland Security, *Privacy Impact Assessment for the Neptune Pilot*, September 2013, http://www.dhs.gov/sites/default/files/publications/privacy-pia-dhs-wide-neptune-09252013.pdf; "Privacy Impact Assessment for the Cerberus Pilot," November 22, 2013, http://www.dhs.gov/sites/default/files/publications/privacy-pia-dhs-cerberus-nov 2013. pdf.

[69] In the first phase, three databases, from different parts of the agency, are fed into Neptune, where the data is then tagged and sorted. From there, the Department of Homeland Security feeds this tagged data into Cerberus, which operates at the classified level. Here, DHS can compare its unclassified and classified information.

[70] For more information, see the Department of Homeland Security's Privacy Office website, http://www.dhs.gov/privacy, and Office for Civil Rights and Civil Liberties, http://www.dhs.gov/office-civilrights-and-civil-liberties.

[71] Most jurisprudence to date does not consider in their entirety big data technologies by the definition used in this report, but rather many of the advanced technologies, such as GPS trackers, that now play a crucial role in big data applications.

[72] *United States v. Jones*, 132 S.Ct. 945, 958 (2012) (Alito, J., concurring).

[73] Ibid at n.3.

[74] Over 70 cities in the U.S. use gunshot detection technology developed and provided by SST Solutions called ShotSpotter. For more information, please visit www.shotspotter.com.

[75] International Association of Chiefs of Police, *Privacy Impact Assessment Report for the Utilization of License Plate Readers*, September 2009, http://www.theiacp.org/Portals/0/pdfs/LPR_Privacy_Impact_Assessment.pdf.

[76] The National Institute of Justice, the Department of Justice's research, development, and evaluation agency, provides detailed information on the use of predictive policing at law enforcement agencies. For more information, visit www.nij.gov/topics/law-enforcement/strategies/predictive-policing.

[77] Andree G. Ferguson, "Big Data and Predictive Reasonable Suspicion," 163 *University of Pennsylvania Law Review*, April 2014, http://ssrn.com/abstract=2394683.

[78] The application of this particular predictive policing technology emerged out of a series of grants issued by the National Institute of Justice the Chicago Police Department, most recently involving Miles Wernick as technical investigator. For more information, see http://www.nij.gov/topics/law-enforcement/strategies/predictive-policing/Pages/research. aspx.

[79] For more information on government crime prediction using statistical methods, refer to Eric Holder, Mary Lou Leary, and Greg Ridgeway, "Predicting Recidivism Risk: New Tool in Philadelphia Shows Great Promise," *National Criminal Justice Reference Service*, February 2013, https://ncjrs.gov/pdffiles1/nij/240695.pdf.

[80] Controversial aspects of the Chicago pilot's methodology are captured by Jay Stanley, "Chicago Police 'Heat List' Renews Old Fears About Government Flagging and Tagging," *American Civil Liberties Union*, February 2014, https://www.aclu.org/blog/technology-and-

liberty/chicago-police-heat-list-renews-old-fearsabout-government-flagging-and; Whet Moser, The Small Social Networks at the Heart of Chicago Violence," *Chicago Magazine*, December 9, 2013, http://www.chicagomag.com/city-life/December-2013/The-SmallSocial-Networks-at-the-Heart-of-Chicago-Violence.

[81] Though some argue big data analysis is merely a new way to expand the scope of what can be considered "suspicion," the program in question uses an algorithmic calculation heavily reliant on an individual's associations without other criminal pretext.

[82] *Katz v. United States*, 389 U.S. 347, 351-52 (1967).

[83] *United States v. Miller*, 425 U.S. 435, 443 (1976).

[84] *Smith v. Maryland*, 442 U.S. 735, 743-44 (1979).

[85] Fred Cate and C. Ben Dutton, "Comments to the 60-Day Cybersecurity Review," *Center for Applied Cybersecurity Research*, March 2009, http://www.whitehouse.gov/files/documents/cyber/Center%20for%20Applied%20Cybersecurity%20Research%20-%20Cybersecurity%20Comments.Cate.pdf; Randy Reitman, "Deep Dive: Updating the Electronic Communications Privacy Act," *Electronic Frontier Foundation*, December 2012, https://www.eff.org/deeplinks/2012/12/deep-dive-updating-electronic-communications-privacy-act.

[86] *United States v. Warshak*, 631 F.3d 266 (6th Cir. 2010).

[87] This assertion was not part of the Supreme Court's holding, but emphasizes the emerging discussion of third-party doctrine. *United States v. Jones*, 132 S.Ct. 945, 957 (2012) (Sotomayor, J., concurring).

[88] The doctrine has been adapted and applied to cell-site location information multiple times, most recently by the Fifth Circuit in *In re Application of the United States for Historical Cell Site Data*, 724 F.3d 600 (5th Cir. 2013) (finding cell site data may be obtained without a probable cause warrant); *United States v. Norris*, No. 2:11-CR-00188-KJM, 2013 WL 4737197 (E.D. Cal. Sept. 3, 2013) (finding defendant who hacked a private wireless network had no reasonable expectation of privacy in his transmissions over that network). Moreover, leading commentators have argued for the continuing vitality of the third-party doctrine in the modern era, including Professor Orin Kerr in Orin S. Kerr, "The Case for the Third-Party Doctrine," 107 *Michigan Law Review* 561 (2009), and Orin S. Kerr, "Defending the Third-Party Doctrine: A Response to Epstein and Murphy," 24 *Berkeley Technology Law Journal* 1229 (2009). *See also United States v. Perrine*, 518 F.3d 1196, 1204 (10th Cir. 2008); *United States v. Forrester*, 512 F.3d 500, 510 (9th Cir. 2008).

[89] See Robert Gellman and Pam Dixon, "Data Brokers and the Federal Government: A New Front in the Battle for Privacy Opens," *World Privacy Forum Report*, Oct. 30, 2013; Chris Hoofnagle, "Big Brother's Little Helpers: How Choicepoint and Other Commercial Data Brokers Collect, Process, and Package Your Data for Law Enforcement," 29 *North Carolina Journal of International Law and Commercial Regulation* 595 (2003); Jon Michaels, "All the President's Spies: Private-Public Intelligence Partnerships in the War on Terror," 96 *California Law Review* 901 (2008).

[90] Office of Management and Budget memorandum M-13-20, *Protecting Privacy while Reducing Improper Payments with the Do Not Pay Initiative* (Aug. 13, 2013), http://www.whitehouse.gov/sites/default/files/omb/memoranda/2013/m-13-20.pdf.

[91] Department of Defense, *Security From Within: A Report of the Independent Review of the Washington Navy Yard Shooting*, November 2013, http://www.defense.gov/pubs/Independent-Review-of-the-WNYShooting-14-Nov-2013.pdf; Department of Defense, Under Secretary of Defense for Intelligence, *Internal Review of the Washington Navy Yard Shooting: A Report to the Secretary of Defense*, November, 2013, http://www.defense.gov/pubs/DoD-Internal-Review-of-the-WNY-Shooting-20-Nov-2013.pdf.

[92] Ibid.

[93] Performance Accountability Council, *Suitability and Security Processes Review, Report to the President*, February 2014, http://www.whitehouse.gov/sites/default/files/omb/reports/suitability-and-security-processreview-report.pdf.

[94] Ford and MacArthur Foundation, *A Future or Failure?: The Flow of Technology Talent into Government and Civil Society*, December 2013, http://www.fordfoundation.org/pdfs/news/afutureoffailure.pdf.

[95] Dan Vesset and Henry Morris, *Unlocking the Business Value of Big Data: Infosys BigDataEdge*, IDC, 2013, http://www.infosys.com/bigdataedge/resources/Documents/unlocking-business-value.pdf

[96] Ibid.

[97] Centre for Information Policy Leadership, *Big Data and Analytics: Seeking Foundations for Effective Privacy Guidance*, February 2013, p. 3-4, http://www.hunton.com/files/Uploads/Documents/News_files/Big_Data_and_Analytics_February_2013.pdf.

[98] FDIC, *2011 FDIC National Survey of Unbanked and Underbanked Households*, 2012, http://www.fdic.gov/householdsurvey/2012_unbankedreport_execsumm.pdf.

[99] John Deighton and Leora Kornfeld, *Economic Value of the Advertising-Supported Internet Ecosystem*, Interactive Advertising Bureau, 2012, http://www.iab.net/economicvalue.

[100] J. Howard Beales and Jeffrey Eisenach, *An Empirical Analysis Of The Value Of Information Sharing in the Market for Online Content*, Navigant Economics, 2014, https://www.aboutads.info/resource/fullvalueinfostudy.pdf.

[101] LUMA Partners, "Display Lumascape," http://www.lumapartners.com/lumascapes/display-ad-tech lumascape.

[102] For information about the industry's opt-out program, see http://www.youradchoices.com/..

[103] See Google Ads Settings at http://www.google.com/settings/ads; Yahoo! Ads Interest Manager at https://info.yahoo.com/privacy/us/yahoo/opt_out/targeting/; Microsoft at http://choice.microsoft.com/en-us/opt-out.

[104] See, e.g., Pedro Leon, Blase Ur, Richard Shay, Yang Wang, Rebecca Balebako, and Lorrie Cranor, "Why Johnny Can't Opt Out: a Usability Evaluation of Tools to Limit Online Behavioral Advertising," Proceedings of the SIGCHI Conference on Human Factors in Computing Systems, 2012, http://dl.acm.org/citation.cfm?doid=2207676.2207759.

[105] See, e.g., Sarah Kidner, "Privacy Fatigue Hits Facebook: Have You Updated Your Settings?," *Which? Conversation*, Oct. 18, 2011, http://conversation.which.co.uk/technology/facebook-privacy-settings-privacyfatigue/; Aleecia McDonald and Lorrie Cranor, "The Cost of Reading Privacy Policies," *4 Information Society: A Journal of Law and Policy for the Information Society*, 543, 544, 564 (2008).

[106] Federal Trade Commission, "A Summary of Your Rights Under the Fair Credit Reporting Act," http://www.consumer.ftc.gov/articles/pdf-0096-fair-credit-reporting-act.pdf.

[107] U.S. Senate Committee on Commerce, Science & Transportation, Majority Staff, "A Review of the Data Broker Industry: Collection, Use, and Sale of Consumer Data for Marketing Purposes," p. ii, December 18, 2013.

[108] Ibid at 22.

[109] Federal Trade Commission, *Protecting Consumer Privacy in an Era of Rapid Change: Recommendations for Business and Consumers*, 2012, http://www.ftc.gov/reports/protecting-consumer-privacy-era-rapid-change-recommendations-businesses-policymakers.

[110] Frank Pasquale, *The Black Box Society: The Secret Algorithm Behind Money and Information*, (Harvard University Press, 2014).

[111] Pam Dixon and Robert Gellman, "The Scoring of America: How Secret Consumer Scores Threaten Your Privacy and Your Future," *World Privacy Forum*, April 2014, http://www.worldprivacyforum.org/wpcontent/uploads/2014/04/WPF_Scoring_of_America_April2014_fs.pdf.

[112] The Government Accounting Office conducted a gap analysis of privacy laws and regulation in its September 2013 report on Information Resellers. See GAO, *Information Resellers: Consumer Privacy Framework Needs to Reflect Changes in Technology and the Marketplace*, GAO-13-663, 2013, http://www.gao.gov/assets/660/658151.pdf.

[113] The Leadership Conference on Civil and Human Rights, "Civil Rights Principles for the Era of Big Data," http://www.civilrights.org/press/2014/civil-rights-principles-big-data.html.

[114] Jennifer Valentino-Devries and Jeremy Singer-Vine, "Websites Vary Prices, Deals Based on Users' Information," *The Wall Street Journal*, December 24, 2012, http://online.wsj.com/news/articles/SB10001424127887323777204578189391813881534.

[115] See New Urban Mechanics, http://www.newurbanmechanics.org/. All information about Street Bump comes from its former project manager James Solomon, who was interviewed by officials from the office of the White House Chief Technology Officer.

[116] Westat Corporation, *Findings of the E-Verify Program Evaluation*, December 2009, Report Submitted to Department of Homeland Security, http://www.uscis.gov/sites/default/files/USCIS/E-Verify/EVerify/Final%20E-Verify%20Report%2012-16-09_2.pdf.

[117] Westat Corporation. *Evaluation of the Accuracy of E-Verify Findings*, July 2012, Report Submitted to Department of Homeland Security, http://www.uscis.gov/sites/default/files/USCIS/Verification/E-Verify/EVerify_Native_Documents/Everify%20Studies/Evaluation%20of%20the%20Accuracy%20of%20EVerify%20Findi ngs.pdf.

[118] Stephen Wicker and Robert Thomas, "A Privacy-Aware Architecture for Demand Response Systems," *44th Hawaii International Conference on System Sciences*, January 2011, http://ieeexplore.ieee.org/xpl/login.jsp?tp=&arnumber=5718673&url=http%3A%2F%2Fieeexplore.ieee.org%2Fxpls%2Fabs_all.jsp%3Farnumber%3D5718673; National Institute of Standards and Technology, *Guidelines for Smart Grid Cyber Security: Vol. 2, Privacy and the Smart Grid*, 2010, http://csrc.nist.gov/publications/nistir/ir7628/nistir-7628_vol2.pdf.

[119] President's Council of Advisors on Science & Technology, *Big Data and Privacy: A Technological Perspective*, The White House, May 1, 2014, whitehouse.gov/bigdata.

[120] Ibid at 36.

[121] Networking and Information Technology Research and Development, *Report on Privacy Research within NITRD*, April 2014, http://www.nitrd.gov/Pubs/Report_on_Privacy_Research_within_NITRD.pdf.

[122] President's Council of Advisors on Science & Technology, *Big Data and Privacy: A Technological Perspective*, The White House, May 1, 2014, p. 20, whitehouse.gov/bigdata.

[123] Craig Mundie, "Privacy Pragmatism: Focus on Data Use, Not Data Collection," *Foreign Affairs*, March/April, 2014, http://www.foreignaffairs.com/articles/140741/craig-mundie/privacy-pragmatism.

In: Big Data
Editor: Cody Agnellutti

ISBN: 978-1-63321-397-5
© 2014 Nova Science Publishers, Inc.

Chapter 2

BIG DATA AND PRIVACY: A TECHNOLOGICAL PERSPECTIVE[*]

President's Council of Advisors on Science and Technology

EXECUTIVE SUMMARY

The ubiquity of computing and electronic communication technologies has led to the exponential growth of data from both digital and analog sources. New capabilities to gather, analyze, disseminate, and preserve vast quantities of data raise new concerns about the nature of privacy and the means by which individual privacy might be compromised or protected.

After providing an overview of this report and its origins, Section 1 describes the changing nature of privacy as computing technology has advanced and big data has come to the fore. The term privacy encompasses not only the famous "right to be left alone," or keeping one's personal matters and relationships secret, but also the ability to share information selectively but not publicly. Anonymity overlaps with privacy, but the two are not identical. Likewise, the ability to make intimate personal decisions without government interference is considered to be a privacy right, as is protection from

---

[*] This is an edited, reformatted and augmented version of a Report to the President, issued by the Executive Office of the President, May 2014.

discrimination on the basis of certain personal characteristics (such as race, gender, or genome). Privacy is not just about secrets.

Conflicts between privacy and new technology have occurred throughout American history. Concern with the rise of mass media such as newspapers in the 19th century led to legal protections against the harms or adverse consequences of "intrusion upon seclusion," public disclosure of private facts, and unauthorized use of name or likeness in commerce. Wire and radio communications led to 20th century laws against wiretapping and the interception of private communications – laws that, PCAST notes, have not always kept pace with the technological realities of today's digital communications.

Past conflicts between privacy and new technology have generally related to what is now termed "small data," the collection and use of data sets by private- and public-sector organizations where the data are disseminated in their original form or analyzed by conventional statistical methods. Today's concerns about big data reflect both the substantial increases in the amount of data being collected and associated changes, both actual and potential, in how they are used.

Big data is big in two different senses. It is big in the quantity and variety of data that are available to be processed. And, it is big in the scale of analysis (termed "analytics") that can be applied to those data, ultimately to make inferences and draw conclusions. By data mining and other kinds of analytics, non- obvious and sometimes private information can be derived from data that, at the time of their collection, seemed to raise no, or only manageable, privacy issues. Such new information, used appropriately, may often bring benefits to individuals and society – Section 2 of this report gives many such examples, and additional examples are scattered throughout the rest of the text. Even in principle, however, one can never know what information may later be extracted from any particular collection of big data, both because that information may result only from the combination of seemingly unrelated data sets, and because the algorithm for revealing the new information may not even have been invented at the time of collection.

The same data and analytics that provide benefits to individuals and society if used appropriately can also create potential harms – threats to individual privacy according to privacy norms both widely shared and personal. For example, large-scale analysis of research on disease, together with health data from electronic medical records and genomic information, might lead to better and timelier treatment for individuals but also to inappropriate disqualification for insurance or jobs. GPS tracking of individuals might lead to better community-based public transportation facilities, but also to inappropriate use of the whereabouts of individuals. A list of the kinds of adverse consequences or harms from which individuals should be protected is proposed in Section 1.4. PCAST believes strongly that the positive benefits of big-data technology are (or can be) greater than any new harms.

Section 3 of the report describes the many new ways in which personal data are acquired, both from original sources, and through subsequent processing. Today, although they may not be aware of it, individuals constantly emit into the environment information whose use or misuse may be a source of privacy concerns. Physically, these information emanations are of two types, which can be called "born digital" and "born analog."

When information is "born digital," it is created, by us or by a computer surrogate, specifically for use by a computer or data processing system. When data are born digital, privacy concerns can arise from over-collection. Over-collection occurs when a program's design intentionally, and sometimes clandestinely, collects information unrelated to its stated purpose. Over-collection can, in principle, be recognized at the time of collection.

When information is "born analog," it arises from the characteristics of the physical world. Such information becomes accessible electronically when it impinges on a sensor such as a camera, microphone, or other engineered device. When data are born analog, they are likely to contain more information than the minimum necessary for their immediate purpose, and for valid reasons. One reason is for robustness of the desired "signal" in the presence of variable "noise." Another is technological convergence, the increasing use of standardized components (e.g., cell-phone cameras) in new products (e.g., home alarm systems capable of responding to gesture).

Data fusion occurs when data from different sources are brought into contact and new facts emerge (see Section 3.2.2). Individually, each data source may have a specific, limited purpose. Their combination, however, may uncover new meanings. In particular, data fusion can result in the identification of individual people, the creation of profiles of an individual, and the tracking of an individual's activities. More broadly, data analytics discovers patterns and correlations in large corpuses of data, using increasingly powerful statistical algorithms. If those data include personal data, the inferences flowing from data analytics may then be mapped back to inferences, both certain and uncertain, about individuals.

Because of data fusion, privacy concerns may not necessarily be recognizable in born-digital data when they are collected. Because of signal-processing robustness and standardization, the same is true of born-analog data – even data from a single source (e.g., a single security camera). Born-digital and born-analog data can both be combined with data fusion, and new kinds of data can be generated from data analytics. The beneficial uses of near-ubiquitous data collection are large, and they fuel an increasingly important set of economic activities. Taken together, these considerations suggest that a policy focus on limiting data collection will not be a broadly applicable or scalable strategy – nor one likely to achieve the right balance between beneficial results and unintended negative consequences (such as inhibiting economic growth).

If collection cannot, in most cases, be limited practically, then what? Section 4 discusses in detail a number of technologies that have been used in the past for privacy protection, and others that may, to a greater or lesser extent, serve as technology building blocks for future policies.

Some technology building blocks (for example, cybersecurity standards, technologies related to encryption, and formal systems of auditable access control) are already being utilized and need to be encouraged in the marketplace. On the other hand, some techniques for privacy protection that have seemed encouraging in the past are useful as supplementary ways to reduce privacy risk, but do not now seem sufficiently robust to be a dependable basis for privacy protection where big data is concerned. For a variety of reasons, PCAST judges anonymization, data deletion, and distinguishing data from metadata (defined below) to be in this category. The framework of notice and consent is also becoming unworkable as a useful foundation for policy.

Anonymization is increasingly easily defeated by the very techniques that are being developed for many legitimate applications of big data. In general, as the size and diversity of available data grows, the likelihood of being able to re-identify individuals (that is, re-associate their records with their names) grows substantially. While anonymization may remain somewhat useful as an added safeguard in some situations, approaches that deem it, by itself, a sufficient safeguard need updating.

While it is good business practice that data of all kinds should be deleted when they are no longer of value, economic or social value often can be obtained from applying big data techniques to masses of data that were otherwise considered to be worthless. Similarly, archival data may also be important to future historians, or for later longitudinal analysis by academic researchers and others. As described above, many sources of data contain latent information about individuals, information that can be known only if the holder expends analytic resources, or that may become knowable only in the future with the development of new data-mining algorithms. In such cases it is practically impossible for the data holder even to surface "all the data about an individual," much less delete it on any specified schedule or in response to an individual's request. Today, given the distributed and redundant nature of data storage, it is not even clear that data, even small data, *can* be destroyed with any high degree of assurance.

As data sets become more complex, so do the attached metadata. Metadata are ancillary data that describe properties of the data such as the time the data were created, the device on which they were created, or the destination of a message. Included in the data or metadata may be identifying information of many kinds. It cannot today generally be asserted that metadata raise fewer privacy concerns than data.

Notice and consent is the practice of requiring individuals to give positive consent to the personal data collection practices of each individual app, program, or web service. Only in some fantasy world do users actually read these notices and understand their implications before clicking to indicate their consent.

The conceptual problem with notice and consent is that it fundamentally places the burden of privacy protection on the individual. Notice and consent creates a non-level playing field in the implicit privacy negotiation between provider and user. The provider offers a complex, take-it-or-leave-it set of terms, while the user, in practice, can allocate only a few seconds to evaluating the offer. This is a kind of market failure.

PCAST believes that the responsibility for using personal data in accordance with the user's preferences should rest with the provider rather than with the user. As a practical matter, in the private sector, third parties chosen by the consumer (e.g., consumer-protection organizations, or large app stores) could intermediate: A consumer might choose one of several "privacy protection profiles" offered by the intermediary, which in turn would vet apps against these profiles. By vetting apps, the intermediaries would create a marketplace for the negotiation of community standards for privacy. The Federal government could encourage the development of standards for electronic interfaces between the intermediaries and the app developers and vendors.

After data are collected, data analytics come into play and may generate an increasing fraction of privacy issues. Analysis, per se, does not directly touch the individual (it is neither collection nor, without additional action, use) and may have no external visibility. By contrast, it is the *use* of a product of analysis, whether in commerce, by government, by the press, or by individuals, that can cause adverse consequences to individuals.

More broadly, PCAST believes that it is the use of data (including born-digital or born-analog data and the products of data fusion and analysis) that is the locus where consequences are produced. This locus is the technically most feasible place to protect privacy. Technologies are emerging, both in the research community and in the commercial world, to describe privacy policies, to record the origins (provenance) of data, their access, and their further use by programs, including analytics, and to determine whether those uses conform to privacy policies. Some approaches are already in practical use.

Given the statistical nature of data analytics, there is uncertainty that discovered properties of groups apply to a particular individual in the group. Making incorrect conclusions about individuals may have adverse consequences for them and may affect members of certain groups disproportionately (e.g., the poor, the elderly, or minorities). Among the technical mechanisms that can be incorporated in a use-based approach are methods for imposing standards for data accuracy and integrity and policies for incorporating useable interfaces that allow an individual to correct the record with voluntary additional information.

PCAST's charge for this study did not ask it to recommend specific privacy policies, but rather to make a relative assessment of the technical feasibilities of different broad policy approaches. Section 5, accordingly, discusses the implications of current and emerging technologies for government policies for privacy protection. The use of technical measures for enforcing privacy can be stimulated by reputational pressure, but such measures are most effective when there are regulations and laws with civil or criminal penalties. Rules and regulations provide both deterrence of harmful actions and incentives to deploy privacy-protecting technologies. Privacy protection cannot be achieved by technical measures alone.

This discussion leads to five recommendations.

Recommendation 1. Policy attention should focus more on the actual uses of big data and less on its collection and analysis.

By actual uses, we mean the specific events where something happens that can cause an adverse consequence or harm to an individual or class of individuals. In the context of big data, these events ("uses") are almost always actions of a computer program or app interacting either with the raw data or with the fruits of analysis of those data. In this formulation, it is not the data themselves that cause the harm, nor the program itself (absent any data), but the confluence of the two. These "use" events (in commerce, by government, or by individuals) embody the necessary specificity to be the subject of regulation. By contrast, PCAST judges that policies focused on the regulation of data collection, storage, retention, a priori limitations on applications, and analysis (absent identifiable actual uses of the data or products of analysis) are unlikely to yield effective strategies for improving privacy. Such policies would be unlikely to be scalable over time, or to be enforceable by other than severe and economically damaging measures.

Recommendation 2. Policies and regulation, at all levels of government, should not embed particular technological solutions, but rather should be stated in terms of intended outcomes.

To avoid falling behind the technology, it is essential that policy concerning privacy protection should address the purpose (the "what") rather than prescribing the mechanism (the "how").

Recommendation 3. With coordination and encouragement from OSTP,[1] the NITRD agencies[2] should strengthen U.S. research in privacy-related technologies and in the relevant areas of social science that inform the successful application of those technologies.

Some of the technology for controlling uses already exists. However, research (and funding for it) is needed in the technologies that help to protect privacy, in the social mechanisms that influence privacy- preserving behavior, and in the legal options that are robust to changes in technology and create appropriate balance among economic opportunity, national priorities, and privacy protection.

Recommendation 4. OSTP, together with the appropriate educational institutions and professional societies, should encourage increased education and training opportunities concerning privacy protection, including career paths for professionals.

Programs that provide education leading to privacy expertise (akin to what is being done for security expertise) are essential and need encouragement. One might envision careers for digital-privacy experts both on the software development side and on the technical management side.

Recommendation 5. The United States should take the lead both in the international arena and at home by adopting policies that stimulate the use of practical privacy-protecting technologies that exist today. It can

exhibit leadership both by its convening power (for instance, by promoting the creation and adoption of standards) and also by its own procurement practices (such as its own use of privacy-preserving cloud services).

PCAST is not aware of more effective innovation or strategies being developed abroad; rather, some countries seem inclined to pursue what PCAST believes to be blind alleys. This circumstance offers an opportunity for U.S. technical leadership in privacy in the international arena, an opportunity that should be taken.

1. Introduction

In a widely noted speech on January 17, 2014, President Barack Obama charged his Counselor, John Podesta, with leading a comprehensive review of big data and privacy, one that would "reach out to privacy experts, technologists, and business leaders and look at how the challenges inherent in big data are being confronted by both the public and private sectors; whether we can forge international norms on how to manage this data; and how we can continue to promote the free flow of information in ways that are consistent with both privacy and security."[3] The President and Counselor Podesta asked the President's Council of Advisors on Science and Technology (PCAST) to assist with the technology dimensions of the review.

For this task PCAST's statement of work reads, in part,

> PCAST will study the technological aspects of the intersection of big data with individual privacy, in relation to both the current state and possible future states of the relevant technological capabilities and associated privacy concerns.
>
> Relevant big data include data and metadata collected, or potentially collectable, from or about individuals by entities that include the government, the private sector, and other individuals. It includes both proprietary and open data, and also data about individuals collected incidentally or accidentally in the course of other activities (e.g., environmental monitoring or the "Internet of Things").

This is a tall order, especially on the ambitious timescale requested by the President. The literature and public discussion of big data and privacy are vast, with new ideas and insights generated daily from a variety of constituencies: technologists in industry and academia, privacy and consumer advocates, legal scholars, and journalists (among others). Independently of PCAST, but informing this report, the Podesta study sponsored three public workshops at universities across the country. Limiting this report's charge to technological, not policy, aspects of the problem narrows PCAST's mandate somewhat, but this is a subject where technology and policy are difficult to separate. In any case, it is the nature of the subject that this report must be regarded as based on a momentary snapshot of the technology, although we believe the key conclusions and recommendations have lasting value.

1.1. Context and Outline of This Report

The ubiquity of computing and electronic communication technologies has led to the exponential growth of online data, from both digital and analog sources. New technological capabilities to create, analyze, and disseminate vast quantities of data raise new concerns about the nature of privacy and the means by which individual privacy might be compromised or protected.

This report discusses present and future technologies concerning this so-called "big data" as it relates to privacy concerns. It is not a complete summary of the technology concerning big data, nor a complete summary of the ways in which technology affects privacy, but focuses on the ways in which big-data and privacy interact. As an example, if Leslie confides a secret to Chris and Chris broadcasts that secret by email or texting, that might be a privacy-infringing use of information technology, but it is not a big-data issue. As another example, if oceanographic data are collected in large quantities by remote sensing, that is big data, but not, in the first instance, a privacy concern. Some data are more privacy-sensitive than others, for example, personal medical data, as distinct from personal data publicly shared by the same individual. Different technologies and policies will apply to different classes of data.

The notions of big data and the notions of individual privacy used in this report are intentionally broad and inclusive. Business consultants Gartner, Inc. define big data as "high-volume, high-velocity and high-variety information assets that demand cost-effective, innovative forms of information processing for enhanced insight and decision making,"[4] while computer scientists reviewing multiple definitions offer the more technical, "a term describing the storage and analysis of large and/or complex data sets using a series of techniques including, but not limited to, NoSQL, MapReduce, and machine learning."[5] (See Sections 3.2.1 and 3.3.1 for discussion of these technical terms.) In a privacy context, the term "big data" typically means data about one or a group of individuals, or that might be analyzed to make inferences about individuals. It might include data or metadata collected by government, by the private sector, or by individuals. The data and metadata might be proprietary or open, they might be collected intentionally or incidentally or accidentally. They might be text, audio, video, sensor-based, or some combination. They might be data collected directly from some source, or data derived by some process of analysis. They might be saved for a long period of time, or they might be analyzed and discarded as they are streamed. In this report, PCAST usually does not distinguish between "data" and "information."

The term "privacy" encompasses not only avoiding observation, or keeping one's personal matters and relationships secret, but also the ability to share information selectively but not publicly. Anonymity overlaps with privacy, but the two are not identical. Voting is recognized as private, but not anonymous, while authorship of a political tract may be anonymous, but it is not private. Likewise, the ability to make intimate personal decisions without government interference is considered to be a privacy right, as is protection from discrimination on the basis of certain personal characteristics (such as an individual's race, gender, or genome). So, privacy is not just about secrets.

The promise of big-data collection and analysis is that the derived data can be used for purposes that benefit both individuals and society. Threats to privacy stem from the deliberate or inadvertent disclosure of collected or derived individual data, the misuse of the data, and the fact that derived data may be inaccurate or false. The technologies that address the confluence of these issues are the subject of this report.[6]

The remainder of this introductory section gives further context in the form of a summary of how the legal concept of privacy developed historically in the United States. Interestingly, and relevant to this report, privacy rights and the development of new technologies have long been intertwined. Today's issues are no exception.

Section 2 of this report is devoted to scenarios and examples, some from today, but most anticipating a near tomorrow. Yogi Berra's much-quoted remark – "It's tough to make predictions, especially about the future" – is germane. But it is equally true for this subject that policies based on out-of-date examples and scenarios are doomed to failure. Big-data technologies are advancing so rapidly that predictions about the future, however imperfect, must guide today's policy development.

Section 3 examines the technology dimensions of the two great pillars of big data: collection and analysis. In a certain sense big data is exactly the confluence of these two: big collection meets big analysis (often termed "analytics"). The technical infrastructure of large-scale networking and computing that enables "big" is also discussed.

Section 4 looks at technologies and strategies for the protection of privacy. Although technology may be part of the problem, it must also be part of the solution. Many current and foreseeable technologies can enhance privacy, and there are many additional promising avenues of research.

Section 5, drawing on the previous sections, contains PCAST's perspectives and conclusions. While it is not within this report's charge to recommend specific policies, it is clear that certain kinds of policies are technically more feasible and less likely to be rendered irrelevant or unworkable by new technologies than others. These approaches are highlighted, along with comments on the technical deficiencies of some other approaches. This section also contains PCAST's recommendations in areas that lie within our charge, that is, other than policy.

1.2. Technology Has Long Driven the Meaning of Privacy

The conflict between privacy and new technology is not new, except perhaps now in its greater scope, degree of intimacy, and pervasiveness. For more than two centuries, values and expectations relating to privacy have been continually reinterpreted and rearticulated in light of the impact of new technologies.

The nationwide postal system advocated by Benjamin Franklin and established in 1775 was a new technology designed to promote interstate commerce. But mail was routinely and opportunistically opened in transit until Congress made this action illegal in 1782. While the Constitution's Fourth Amendment codified the heightened privacy protection afforded to people in their homes or on their persons (previously principles of British common law), it took another century of technological challenges to expand the concept of privacy rights into more abstract spaces, including the electronic. The invention of the telegraph and, later, telephone created new tensions that were slow to be resolved. A bill to protect the privacy of telegrams, introduced in Congress in 1880, was never passed.[7]

It was not telecommunications, however, but the invention of the portable, consumer-operable camera (soon known as the Kodak) that gave impetus to Warren and Brandeis's 1890 article "The Right to Privacy,"[8] then a controversial title, but now viewed as the foundational document for modern privacy law. In the article, Warren and Brandeis gave voice to the concern that "[i]nstantaneous photographs and newspaper enterprise have invaded the sacred precincts of private and domestic life; and numerous mechanical devices threaten to make good the prediction that 'what is whispered in the closet shall be proclaimed from the house-tops,'" further noting that "[f]or years there has been a feeling that the law must afford some remedy for the unauthorized circulation of portraits of private persons..."[9]

Warren and Brandeis sought to articulate the right of privacy between individuals (whose foundation lies in civil tort law). Today, many states recognize a number of privacy-related harms as causes for civil or criminal legal action (further discussed in Section 1.4).[10]

From Warren and Brandeis' "right to privacy," it took another 75 years for the Supreme Court to find, in *Griswold v. Connecticut*[11] (1965), a right to privacy in the "penumbras" and "emanations" of other constitutional protections (as Justice William O. Douglas put it, writing for the majority).[12] With a broad perspective, scholars today recognize a number of different legal meanings for "privacy." Five of these seem particularly relevant to this PCAST report:

(1) The individual's right to keep secrets or seek seclusion (the famous "right to be left alone" of Brandeis' 1928 dissenting opinion in *Olmstead v. United States*).[13]
(2) The right to anonymous expression, especially (but not only) in political speech (as in *McIntyre v. Ohio Elections Commission*[14])
(3) The ability to control access by others to personal information after it leaves one's exclusive possession (for example, as articulated in the FTC's Fair Information Practice Principles).[15]
(4) The barring of some kinds of negative consequences from the use of an individual's personal information (for example, job discrimination on the basis of personal DNA, forbidden in 2008 by the Genetic Information Nondiscrimination Act[16]).
(5) The right of the individual to make intimate decisions without government interference, as in the domains of health, reproduction, and sexuality (as in *Griswold*).

These are asserted, not absolute, rights. All are supported, but also circumscribed, by both statute and case law. With the exception of number 5 on the list (a right of "decisional privacy" as distinct from "informational privacy"), all are applicable in varying degrees both to citizen-government interactions and to citizen-citizen interactions. Collisions between new technologies and privacy rights have occurred in all five. A patchwork of state and federal laws have addressed concerns in many sectors, but to date there has not been comprehensive legislation to handle these issues. Collisions between new technologies and privacy rights should be expected to continue to occur.

1.3. What Is Different Today?

New collisions between technologies and privacy have become evident, as new technological capabilities have emerged at a rapid pace. It is no longer clear that the five privacy concerns raised above, or their current legal interpretations, are sufficient in the court of public opinion.

Much of the public's concern is with the harm done by the use of personal data, both in isolation or in combination. Controlling access to personal data after they leave one's exclusive possession has been seen historically as a means of controlling potential harm. But today, personal data may never be, or have been, within one's possession – for instance they may be acquired passively from external sources such as public cameras and sensors, or without one's knowledge from public electronic disclosures by others using social media. In addition, personal data may be derived from powerful data analyses (see Section 3.2) whose use and output is unknown to the individual. Those analyses sometimes yield valid conclusions that the individual would not want disclosed. Worse yet, the analyses can produce false positives or false negatives -- information that is a consequence of the analysis but is not true or correct. Furthermore, to a much greater extent than before, the same personal data have both beneficial and harmful uses, depending on the purposes for which and the contexts in which they are used. Information supplied by the individual might be used only to derive other information such as identity or a correlation, after which it is not needed. The derived data, which were never under the individual's control, might then be used either for good or ill.

In the current discourse, some assert that the issues concerning privacy protection are collective as well as individual, particularly in the domain of civil rights – for example, identification of certain individuals at a gathering using facial recognition from videos, and the inference that other individuals at the same gathering, also identified from videos, have similar opinions or behaviors.

Current circumstances also raise issues of how the right to privacy extends to the public square, or to quasi-private gatherings such as parties or classrooms. If the observers in these venues are not just people, but also both visible and invisible recording devices with enormous fidelity and easy paths to electronic promulgation and analysis, does that change the rules?

Also rapidly changing are the distinctions between government and the private sector as potential threats to individual privacy. Government is not just a "giant corporation." It has a monopoly in the use of force; it has no direct competitors who seek market advantage over it and may thus motivate it to correct missteps. Governments have checks and balances, which can contribute to self-imposed limits on what they may do with people's information. Companies decide how they will use such information in the context of such factors as competitive advantages and risks, government regulation, and perceived threats and consequences of lawsuits. It is thus appropriate that there are different sets of constraints on the public and private sectors. But government has a set of authorities – particularly in the areas of law enforcement and national security – that place it in a uniquely powerful position, and therefore the restraints placed on its collection and use of data deserve special attention. Indeed, the need for such attention is heightened because of the increasingly blurry line between public and private data.

While these differences are real, big data is to some extent a leveler of the differences between government and companies. Both governments and companies have potential access to the same sources of data and the same analytic tools. Current rules may allow government to purchase or otherwise obtain data from the private sector that, in some cases, it could not legally collect itself,[17] or to outsource to the private sector analyses it could not itself legally perform.[18] The possibility of government exercising, without proper safeguards, its own monopoly powers and also having unfettered access to the private information marketplace is unsettling.

What kinds of actions should be forbidden both to government (Federal, state, and local, and including law enforcement) and to the private sector? What kinds should be forbidden to one but not the other? It is unclear whether current legal frameworks are sufficiently robust for today's challenges.

1.4. Values, Harms, and Rights

As was seen in Sections 1.2 and 1.3, new privacy rights usually do not come into being as academic abstractions. Rather, they arise when technology encroaches on widely shared values. Where there is consensus on values, there can also be consensus on what kinds of harms to individuals may be an affront to those values. Not all such harms may be preventable or remediable by government actions, but, conversely, it is unlikely that government actions will be welcome or effective if they are not grounded to some degree in values that are widely shared.

In the realm of privacy, Warren and Brandeis in 1890[19] (see Section 1.2) began a dialogue about privacy that led to the evolution of the right in academia and the courts, later crystalized by William Prosser as four distinct harms that had come to earn legal protection.[20] A direct result is that, today, many states recognize as causes for legal action the four harms that Prosser enumerated,[21] and which have become (though varying from state to state[22]) privacy "rights." The harms are:

- Intrusion upon seclusion. A person who intentionally intrudes, physically or otherwise (now including electronically), upon the solitude or seclusion of another person or her private affairs or concerns, can be subject to liability for the invasion of her privacy, but only if the intrusion would be highly offensive to a reasonable person.
- Public disclosure of private facts. Similarly, a person can be sued for publishing private facts about another person, even if those facts are true. Private facts are those about someone's personal life that have not previously been made public, that are not of legitimate public concern, and that would be offensive to a reasonable person.
- "False light" or publicity. Closely related to defamation, this harm results when false facts are widely published about an individual. In some states, false light includes untrue implications, not just untrue facts as such.
- Misappropriation of name or likeness. Individuals have a "right of publicity" to control the use of their name or likeness in commercial settings.

It seems likely that most Americans today continue to share the values implicit in these harms, even if the legal language (by now refined in thousands of court decisions) strikes one as archaic and quaint. However, new technological insults to privacy, actual or prospective, and a century's evolution of social values (for example, today's greater recognition of the rights of minorities, and of rights associated with gender), may require a longer list than sufficed in 1960.

Although PCAST's engagement with this subject is centered on technology, not law, any report on the subject of privacy, including PCAST's, should be grounded in the values of its day. As a starting point for discussion, albeit only a snapshot of the views of one set of technologically minded Americans, PCAST offers some possible augmentations to the established list of harms, each of which suggests a possible underlying right in the age of big data.

PCAST also believes strongly that the positive benefits of technology are (or can be) greater than any new harms. Almost every new harm is related to or "adjacent to" beneficial uses of the same technology.[23] To emphasize this point, for each suggested new harm, we describe a related beneficial use.

- **Invasion of private communications.** Digital communications technologies make social networking possible across the boundaries of geography, and enable social and political participation on previously unimaginable scales. An individual's right to private communication, secured for written mail and wireline telephone in part by the isolation of their delivery infrastructure, may need reaffirmation in the digital era, however, where all kinds of "bits" share the same pipelines, and the barriers to interception are often much lower. (In this context, we discuss the use and limitations of encryption in Section 4.2.)
- **Invasion of privacy in a person's virtual home.** The Fourth Amendment gives special protection against government intrusion into the home, for example the protection of private records within the home; tort law offers protection against similar non-government intrusion. The new "virtual home" includes the Internet, cloud storage, and other services. Personal data in the cloud can be accessible and organized. Photographs and records in the cloud can be shared with family and friends, and can be passed down to future generations. The underlying social value, the "home as one's castle," should logically extend to one's "castle in the cloud," but this protection has not been preserved in the new virtual home. (We discuss this subject further in Section 2.3.)
- **Public disclosure of inferred private facts.** Powerful data analytics may infer personal facts from seemingly harmless input data. Sometimes the inferences are beneficial. At its best, targeted

advertising directs consumers to products that they actually want or need. Inferences about people's health can lead to better and timelier treatments and longer lives. But before the advent of big data, it could be assumed that there was a clear distinction between public and private information: either a fact was "out there" (and could be pointed to), or it was not. Today, analytics may discover facts that are no less private than yesterday's purely private sphere of life. Examples include inferring sexual preference from purchasing patterns, or early Alzheimer's disease from key-click streams. In the latter case, the private fact may not even be known to the individual in question. (Section 3.2 discusses the technology behind the data analytics that makes such inferences possible.) The public disclosure of such information (and possibly also some non-public commercial uses) seems offensive to widely shared values.

- **Tracking, stalking, and violations of locational privacy.** Today's technologies easily determine an individual's current or prior location. Useful location-based services include navigation, suggesting better commuter routes, finding nearby friends, avoiding natural hazards, and advertising the availability of nearby goods and services. Sighting an individual in a public place can hardly be a private fact. When big data allows such sightings, or other kinds of passive or active data collection, to be assembled into the continuous locational track of an individual's private life, however, many Americans (including Supreme Court Justice Sotomayor, for example[24]) perceive a potential affront to a widely accepted "reasonable expectation of privacy."

- **Harm arising from false conclusions about individuals, based on personal profiles from big-data analytics.** The power of big data, and therefore its benefit, is often correlational. In many cases the "harms" from statistical errors are small, for example the incorrect inference of a movie preference; or the suggestion that a health issue be discussed with a physician, following from analyses that may, on average, be beneficial, even when a particular instance turns out to be a false alarm. Even when predictions are statistically valid, moreover, they may be untrue about particular individuals − and mistaken conclusions may cause harm. Society may not be willing to excuse harms caused by the uncertainties inherent in statistically valid algorithms. These harms may unfairly burden particular classes of individuals, for example, racial minorities or the elderly.

- **Foreclosure of individual autonomy or self-determination.** Data analyses about large populations can discover special cases that apply to individuals within that population. For example, by identifying differences in "learning styles," big data may make it possible to personalize education in ways that recognize every individual's potential and optimize that individual's achievement. But the projection of population factors onto individuals can be misused. It is widely accepted that individuals should be able to make their own choices and pursue opportunities that are not necessarily typical, and that no one should be denied the chance to achieve more than some statistical expectation of themselves. It would offend our values if a child's choices in video games were later used for educational tracking (for example, college admissions). Similarly offensive would be a future, akin to Philip K. Dick's science fiction short story adapted by Steven Spielberg in the film *Minority Report*, where "pre-crime" is statistically identified and punished.[25]
- **Loss of anonymity and private association.** Anonymity is not acceptable as an enabler of committing fraud, or bullying, or cyber-stalking, or improper interactions with children. Apart from wrongful behavior, however, the individual's right to choose to be anonymous is a long held American value (as, for example, the anonymous authorship of the Federalist papers). Using data to (re-) identify an individual who wishes to be anonymous (except in the case of legitimate governmental functions, such as law enforcement) is regarded as a harm. Similarly, individuals have a right of private association with groups or other individuals, and the identification of such associations may be a harm.

While in no sense is the above list intended to be complete, it does have a few intentional omissions. For example, individuals may want big data to be used "fairly," in the sense of treating people equally, but (apart from the small number of protected classes already defined by law) it seems impossible to turn this into a right that is specific enough to be meaningful. Likewise, individuals may want the ability to know what others know about them; but that is surely not a right from the pre-digital age; and, in the current era of statistical analysis, it is not so easy to define what "know" means. This important issue is discussed in Section 3.1.2, and again taken up in section 5, where the attempt is to focus on actual harms done by the *use* of information, not by a concept as technically ambiguous as whether information is *known*.

2. EXAMPLES AND SCENARIOS

This section seeks to make Section 1's introductory discussion more concrete by sketching some examples and scenarios. While some of these applications of technology are in use today, others comprise PCAST's technological prognostications about the near future, up to perhaps 10 years from today. Taken together the examples and scenarios are intended to illustrate both the enormous benefits that big data can provide and also the privacy challenges that may accompany these benefits.

In the following three sections, it will be useful to develop some scenarios more completely than others, moving from very brief examples of things happening today to more fully developed scenarios set in the future.

2.1. Things Happening Today or Very Soon

Here are some relevant examples:

- Pioneered more than a decade ago, devices mounted on utility poles are able to sense the radio stations being listened to by passing drivers, with the results sold to advertisers.[26]
- In 2011, automatic license-plate readers were in use by three quarters of local police departments surveyed. Within 5 years, 25% of departments expect to have them installed on all patrol cars, alerting police when a vehicle associated with an outstanding warrant is in view.[27] Meanwhile, civilian uses of license-plate readers are emerging, leveraging cloud platforms and promising multiple ways of using the information collected.[28]
- Experts at the Massachusetts Institute of Technology and the Cambridge Police Department have used a machine-learning algorithm to identify which burglaries likely were committed by the same offender, thus aiding police investigators.[29]
- Differential pricing (offering different prices to different customers for essentially the same goods) has become familiar in domains such as airline tickets and college costs. Big data may increase the power and prevalence of this practice and may also decrease even further its transparency.[30]

- The UK firm FeatureSpace offers machine-learning algorithms to the gaming industry that may detect early signs of gambling addiction or other aberrant behavior among online players.[31]
- Retailers like CVS and AutoZone analyze their customers' shopping patterns to improve the layout of their stores and stock the products their customers want in a particular location.[32] By tracking cell phones, RetailNext offers bricks-and-mortar retailers the chance to recognize returning customers, just as cookies allow them to be recognized by on-line merchants.[33] Similar WiFi tracking technology could detect how many people are in a closed room (and in some cases their identities).
- The retailer Target inferred that a teenage customer was pregnant and, by mailing her coupons intended to be useful, unintentionally disclosed this fact to her father.[34]
- The author of an anonymous book, magazine article, or web posting is frequently "outed" by informal crowd sourcing, fueled by the natural curiosity of many unrelated individuals.[35]
- Social media and public sources of records make it easy for anyone to infer the network of friends and associates of most people who are active on the web, and many who are not.[36]
- Marist College in Poughkeepsie, New York, uses predictive modeling to identify college students who are at risk of dropping out, allowing it to target additional support to those in need.[37]
- The Durkheim Project, funded by the U.S. Department of Defense, analyzes social-media behavior to detect early signs of suicidal thoughts among veterans.[38]
- LendUp, a California-based startup, sought to use nontraditional data sources such as social media to provide credit to underserved individuals. Because of the challenges in ensuring accuracy and fairness, however, they have been unable to proceed.[39,40]
- Insight into the spread of hospital-acquired infections has been gained through the use of large amounts of patient data together with personal information about uninfected patients and clinical staff.[41]
- Individuals' heart rates can be inferred from the subtle changes in their facial coloration that occur with each beat, enabling inferences about their health and emotional state.[42]

2.2. Scenarios of the Near Future in Healthcare and Education

Here are a few examples of the kinds of scenarios that can readily be constructed.

2.2.1. Healthcare: Personalized Medicine

Not all patients who have a particular disease are alike, nor do they respond identically to treatment. Researchers will soon be able to draw on millions of health records (including analog data such as scans in addition to digital data), vast amounts of genomic information, extensive data on successful and unsuccessful clinical trials, hospital records, and so forth. In some cases they will be able to discern that among the diverse manifestations of the disease, a subset of the patients have a collection of traits that together form a variant that responds to a particular treatment regime.

Since the result of the analysis could lead to better outcomes for particular patients, it is desirable to identify those individuals in the cohort, contact them, treat their disease in a novel way, and use their experiences in advancing the research. Their data may have been gathered only anonymously, however, or it may have been de-identified.

Solutions may be provided by specific new technologies for the protection of database privacy. These may create a protected query mechanism so individuals can find out whether they are in the cohort, or provide an alert mechanism based on the cohort characteristics so that, when a medical professional sees a patient in the cohort, a notice is generated.

2.2.2. Healthcare: Detection of Symptoms by Mobile Devices

Many baby boomers wonder how they might detect Alzheimer's disease in themselves. What would be better to observe their behavior than the mobile device that connects them to a personal assistant in the cloud (e.g., Siri or OK Google), helps them navigate, reminds them what words mean, remembers to do things, recalls conversations, measures gait, and otherwise is in a position to detect gradual declines on traditional and novel medical indicators that might be imperceptible even to their spouses?

At the same time, any leak of such information would be a damaging betrayal of trust. What are individuals' protections against such risks? Can the inferred information about individuals' health be sold, without additional consent, to third parties (e.g., pharmaceutical companies)? What if this is a stated condition of use of the app? Should information go to individuals' personal physicians with their initial consent but not a subsequent confirmation?

2.2.3. Education

Drawing on millions of logs of online courses, including both massive open on-line courses (MOOCs) and smaller classes, it will soon be possible to create and maintain longitudinal data about the abilities and learning styles of millions of students. This will include not just broad aggregate information like grades, but fine-grained profiles of how individual students respond to multiple new kinds of teaching techniques, how much help they need to master concepts at various levels of abstraction, what their attention span is in various contexts, and so forth. A MOOC platform can record how long a student watches a particular video; how often a segment is repeated, sped up, or skipped; how well a student does on a quiz; how many times he or she misses a particular problem; and how the student balances watching content to reading a text. As the ability to present different material to different students materializes in the platforms, the possibility of blind, randomized A/B testing enables the gold standard of experimental science to be implemented at large scale in these environments.[43]

Similar data are also becoming available for residential classes, as learning-management systems (such as Canvas, Blackboard, or Desire2Learn) expand their roles to support innovative pedagogy. In many courses one can now get moment-by-moment tracking of the student's engagement with the course materials and correlate that engagement with the desired learning outcomes.

With this information, it will be possible not only to greatly improve education, but also to discover what skills, taught to which individuals at which points in childhood, lead to better adult performance in certain tasks, or to adult personal and economic success. While these data could revolutionize educational research, the privacy issues are complex.[44]

There are many privacy challenges in this vision of the future of education. Knowledge of early performance can create implicit biases[45] that color later instruction and counseling. There is great potential for misuse, ostensibly for the social good, in the massive ability to direct students into high- or low-potential tracks. Parents and others have access to sensitive information about children, but mechanisms rarely exist to change those permissions when the child reaches majority.

2.3. Challenges to the Home's Special Status

The home has special significance as a sanctuary of individual privacy. The Fourth Amendment's list, "persons, houses, papers, and effects," puts only the physical body in the rhetorically more prominent position; and a house is often the physical container for the other three, a boundary inside of which enhanced privacy rights apply.

Existing interpretations of the Fourth Amendment are inadequate for the present world, however. We, along with the "papers and effects" contemplated by the Fourth Amendment, live increasingly in cyberspace, where the physical boundary of the home has little relevance. In 1980, a family's financial records were paper documents, located perhaps in a desk drawer inside the house. By 2000, they were migrating to the hard drive of the home computer – but still within the house. By 2020, it is likely that most such records will be in the cloud, not just outside the house, but likely replicated in multiple legal jurisdictions – because cloud storage typically uses location diversity to achieve reliability. The picture is the same if one substitutes for financial records something like "political books we purchase," or "love letters that we receive," or "erotic videos that we watch." Absent different policy, legislative, and judicial approaches, the physical sanctity of the home's papers and effects is rapidly becoming an empty legal vessel.

The home is also the central locus of Brandeis' "right to be left alone." This right is also increasingly fragile, however. Increasingly, people bring sensors into their homes whose immediate purpose is to provide convenience, safety, and security. Smoke and carbon monoxide alarms are common, and often required by safety codes.[46] Radon detectors are usual in some parts of the country. Integrated air monitors that can detect and identify many different kinds of pollutants and allergens are readily foreseeable. Refrigerators may soon be able to "sniff" for gases released from spoiled food, or, as another possible path, may be able to "read" food expiration dates from radio-frequency identification (RFID) tags in the food's packaging. Rather than today's annoying cacophony of beeps, tomorrow's sensors (as some already do today) will interface to a family through integrated apps on mobile devices or display screens. The data will have been processed and interpreted. Most likely that processing will occur in the cloud. So, to deliver services the consumer wants, much data will need to have left the home.

Environmental sensors that enable new food and air safety may also be able to detect and characterize tobacco or marijuana smoke. Health care or health insurance providers may want assurance that self-declared nonsmokers are telling the truth. Might they, as a condition of lower premiums, require the homeowner's consent for tapping into the environmental monitors' data? If the monitor detects heroin smoking, is an insurance company obligated to report this to the police? Can the insurer cancel the homeowner's property insurance?

To some, it seems farfetched that the typical home will foreseeably acquire cameras and microphones in every room, but that appears to be a likely trend. What can your cell phone (already equipped with front and back cameras) hear or see when it is on the nightstand next to your bed? Tablets, laptops, and many desktop computers have cameras and microphones. Motion detector technology for home intrusion alarms will likely move from ultrasound and infrared to imaging cameras – with the benefit of fewer false alarms and the ability to distinguish pets from people. Facial-recognition technology will allow further security and convenience. For the safety of the elderly, cameras and microphones will be able to detect falls or collapses, or calls for help, and be networked to summon aid.

People naturally communicate by voice and gesture. It is inevitable that people will communicate with their electronic servants in both such modes (necessitating that they have access to cameras and microphones).

Companies such as PrimeSense, an Israeli firm recently bought by Apple,[47] are developing sophisticated computer-vision software for gesture reading, already a key feature in the consumer computer game console market (e.g., Microsoft Kinect). Consumer televisions are already among the first "appliances" to respond to gesture; already, devices such as the Nest smoke detector respond to gestures.[48] The consumer who taps his temple to signal a spoken command to Google Glass[49] may want to use the same gesture for the television, or for that matter for the thermostat or light switch, in any room at home. This implies omnipresent audio and video collection within the home.

All of these audio, video, and sensor data will be generated within the supposed sanctuary of the home. But they are no more likely to stay in the home than the "papers and effects" already discussed. Electronic devices in the home already invisibly communicate to the outside world via multiple separate infrastructures: The cable industry's hardwired connection to the home provides multiple types of two-way communication, including broadband Internet. Wireline phone is still used by some home-intrusion alarms and satellite TV receivers, and as the physical layer for DSL broadband subscribers. Some home devices use the cell-phone wireless infrastructure. Many others piggyback on the home Wi-Fi network that is increasingly a necessity of modern life. Today's smart home-entertainment system knows

Big Data and Privacy: A Technological Perspective

what a person records on a DVR, what she actually watches, and when she watches it. Like personal financial records in 2000, this information today is in part localized inside the home, on the hard drive inside the DVR. As with financial information today, however, it is on track to move into the cloud. Today, Netflix or Amazon can offer entertainment suggestions based on customers' past key-click streams and viewing history on their platforms. Tomorrow, even better suggestions may be enabled by interpreting their minute-by-minute facial expressions as seen by the gesture-reading camera in the television.

These collections of data are benign, in the sense that they are necessary for products and services that consumers will knowingly demand. Their challenges to privacy arise both from the fact that their analog sensors necessarily collect more information than is minimally necessary for their function (see Section 3.1.2), and also because their data practically cry out for secondary uses ranging from innovative new products to marketing bonanzas to criminal exploits. As in many other kinds of big data, there is ambiguity as to data ownership, data rights, and allowed data use. Computer-vision software is likely already able to read the brand labels on products in its field of view – this is a much easier technology than facial recognition. If the camera in your television knows what brand of beer you are drinking while watching a football game, and knows whether you opened the bottle before or after the beer ad, who (if anyone) is allowed to sell this information to the beer company, or to its competitors? Is the camera allowed to read brand names when the television set is supposedly off? Can it watch for magazines or political leaflets? If the RFID tag sensor in your refrigerator usefully detects out-of-date food, can it also report your brand choices to vendors? Is this creepy and strange, or a consumer financial benefit when every supermarket can offer you relevant coupons?[50] Or (the dilemma of differential pricing[51]) is it any different if the data are used to offer *others* a better deal while *you* pay full price because your brand loyalty is known to be strong?

About one-third of Americans rent, rather than own, their residences. This number may increase with time as a result of long-term effects of the 2007 financial crisis, as well as aging of the U.S. population. Today and foreseeably, renters are less affluent, on average, than homeowners. The law demarcates a fine line between the property rights of landlords and the privacy rights of tenants. Landlords have the right to enter their property under various conditions, generally including where the tenant has violated health or safety codes, or to make repairs. As more data are collected within the home, the rights of tenant and landlord may need new adjustment. If environmental monitors are fixtures of the landlord's property, does she have an unconditional right to their data? Can she sell those data? If the lease so provides, can she evict the tenant if the monitor repeatedly detects cigarette smoke, or a camera sensor is able to distinguish a prohibited pet?

If a third party offers facial recognition services for landlords (no doubt with all kinds of cryptographic safeguards!), can the landlord use these data to enforce lease provisions against subletting or additional residents? Can she require such monitoring as a condition of the lease? What if the landlord's cameras are outside the doors, but keep track of everyone who enters or leaves her property? How is this different from the case of a security camera across the street that is owned by the local police?

2.4. Tradeoffs among Privacy, Security, and Convenience

Notions of privacy change generationally. One sees today marked differences between the younger generation of "digital natives" and their parents or grandparents. In turn, the children of today's digital natives will likely have still different attitudes about the flow of their personal information. Raised in a world with digital assistants who know everything about them, and (one may hope) with wise policies in force to govern use of the data, future generations may see little threat in scenarios that individuals today would find threatening, if not Orwellian. PCAST's final scenario, perhaps at the outer limit of its ability to prognosticate, is constructed to illustrate this point.

Taylor Rodriguez prepares for a short business trip. She packed a bag the night before and put it outside the front door of her home for pickup. No worries that it will be stolen: The camera on the streetlight was watching it; and, in any case, almost every item in it has a tiny RFID tag. Any would-be thief would be tracked and arrested within minutes. Nor is there any need to give explicit instructions to the delivery company, because the cloud knows Taylor's itinerary and plans; the bag is picked up overnight and will be in Taylor's destination hotel room by the time of her arrival.

Taylor finishes breakfast and steps out the front door. Knowing the schedule, the cloud has provided a self-driving car, waiting at the curb. At the airport, Taylor walks directly to the gate – no need to go through any security. Nor are there any formalities at the gate: A twenty-minute "open door" interval is provided for passengers to stroll onto the plane and take their seats (which each sees individually highlighted in his or her wearable optical device). There are no boarding passes and no organized lines. Why bother,

Big Data and Privacy: A Technological Perspective 119

when Taylor's identity (as for everyone else who enters the airport) has been tracked and is known absolutely? When her known information emanations (phone, RFID tags in clothes, facial recognition, gait, emotional state) are known to the cloud, vetted, and essentially unforgeable? When, in the unlikely event that Taylor has become deranged and dangerous, many detectable signs would already have been tracked, detected, and acted on?

Indeed, everything that Taylor carries has been screened far more effectively than any rushed airport search today. Friendly cameras in every LED lighting fixture in Taylor's house have watched her dress and pack, as they do every day. Normally these data would be used only by Taylor's personal digital assistants, perhaps to offer reminders or fashion advice. As a condition of using the airport transit system, however, Taylor has authorized the use of the data for ensuring airport security and public safety.

Taylor's world seems creepy to us. Taylor has accepted a different balance among the public goods of convenience, privacy, and security than would most people today. Taylor acts in the unconscious belief (whether justified or not, depending on the nature and effectiveness of policies in force) that the cloud and its robotic servants are trustworthy in matters of personal privacy. In such a world, major improvements in the convenience and security of everyday life become possible.

3. COLLECTION, ANALYTICS, AND SUPPORTING INFRASTRUCTURE

Big data is big in two different senses. It is big in the quantity and variety of data that are available to be processed. And, it is big in the scale of analysis ("analytics") that can be applied to those data, ultimately to make inferences. Both kinds of "big" depend on the existence of a massive and widely available computational infrastructure, one that is increasingly being provided by cloud services. This section expands on these basic concepts.

3.1. Electronic Sources of Personal Data

Since early in the computer age, public and private entities have been assembling digital information about people. Databases of personal information were created during the days of "batch processing."[52] Indeed, early descriptions of database technology often talk about personnel records used for payroll applications. As computing power increased, more and more business applications moved to digital form. There now are digital telephone-call records, credit-card transaction records, bank-account records, email repositories, and so on. As interactive computing has advanced, individuals have entered more and more data about themselves, both for self-identification to an online service and for productivity tools such as financial-management systems.

These digital data are normally accompanied by "metadata" or ancillary data that explain the layout and meaning of the data they describe. Databases have schemas and email has headers,[53] as do network packets.[54] As data sets become more complex, so do the attached metadata. Included in the data or metadata may be identifying information such as account numbers, login names, and passwords. There is no reason to believe that metadata raise fewer privacy concerns than the data they describe.

In recent times, the kinds of electronic data available about people have increased substantially, in part because of the emergence of social media and in part because of the growth in mobile devices, surveillance devices, and a diversity of networked sensors. Today, although they may not be aware of it, individuals constantly emit into the environment information whose use or misuse may be a source of privacy concerns. Physically, these information emanations are of two types, which can be called "born digital" or "born analog."

3.1.1. "Born Digital" Data

When information is "born digital," it is created, by us or by a computer surrogate, specifically for digital use – that is, for use by a computer or data-processing system. Examples of data that are born digital include:

- email and text messaging
- input via mouse-clicks, taps, swipes, or keystrokes on a phone, tablet, computer, or video game; that is, data that people intentionally enter into a device
- GPS location data
- metadata associated with phone calls: the numbers dialed from or to, the time and duration of calls
- data associated with most commercial transactions: credit-card swipes, bar-code reads, reads of RFID tags (as used for anti-theft and inventory control)

Big Data and Privacy: A Technological Perspective 121

- data associated with portal access (key card or ID badge reads) and toll-road access (remote reads of RFID tags)
- metadata that our mobile devices use to stay connected to the network, including device location and status
- increasingly, data from cars, televisions, appliances: the "Internet of Things"

Consumer-tracking data provide an example of born-digital data that has become economically important. It is generally possible for companies to aggregate large amounts of data and then use those data for marketing, advertising, or many other activities. The traditional mechanism has been to use cookies, small data files that a browser can leave on a user's computer (pioneered by Netscape two decades ago). The technique is to leave a cookie when a user first visits a site and then be able to correlate that visit with a subsequent event. This information is very valuable to retailers and forms the basis of many of the advertising businesses of the last decade. There has been a variety of proposals to regulate such tracking,[55] and many countries require opt-in permission before this tracking is done. Cookies involve relatively simple pieces of information that proponents represent as unlikely to be abused. Although not always aware of the process, people accept such tracking in return for a free or subsidized service.[56] At the same time, cookie-free alternatives are sometimes available.[57] Even without cookies, so-called "fingerprinting" techniques can often identify a user's computer or mobile device uniquely by the information that it exposes publicly, such as the size of its screen, its installed fonts, and other features.[58] Most technologists believe that applications will move away from cookies, that cookies are too simple an idea, and that there are better analytics coming and better approaches being invented. The economic incentives for consumer tracking will remain, however, and big data will allow for more precise responses.

Tracking is also the enabling technology of some more nefarious uses. Unfortunately, many social networking apps begin by taking a person's contact list and spamming all the recipients with advertising for the app. This technique is often abused, especially by small start-ups who may assess the value gained by reaching new customers as being greater than the value lost to their reputation for honoring privacy.

All information that is born digital shares certain characteristics. It is created in identifiable units for particular purposes. These units are in most cases "data packets" of one or another standard type. Since they are created by intent, the information that they contain is usually limited, for reasons of efficiency and good engineering design, to support the immediate purpose for which they are collected.

When data are born digital, privacy concerns can arise in two different modes, one obvious ("over-collection"), the other more recent and subtle ("data fusion"). Over-collection occurs when an engineering design intentionally, and sometimes clandestinely, collects information unrelated to its stated purpose. While your smartphone could easily photograph and transmit to a third party your facial expression as you type every keystroke of a text message, or could capture all keystrokes, thereby recording text that you had deleted, these would be inefficient and unreasonable software design choices for the default text-messaging app. In that context they would be instances of over-collection.

A recent example of over-collection was the *Brightest Flashlight Free* phone app, downloaded by more than 50 million users, which passed back to its vendor its location every time the flashlight was used. Not only is location information unnecessary for the illumination function of a flashlight, but it also discloses personal information that the user might wish to keep private. The Federal Trade Commission issued a complaint because the fine print on the notice-and-consent screen (see Section 4.3) had neglected to disclose that location information, whose collection was disclosed, would be sold to third parties, such as advertisers.[59,60] One sees in this example the limitations of the notice-and-consent framework: A more detailed initial fine-print disclosure by *Brightest Flashlight Free*, which almost no one would have actually read, would likely have forestalled any FTC action without much affecting the number of downloads.

In contrast to over-collection, data fusion occurs when data from different sources are brought into contact and new, often unexpected, phenomena emerge (see Section 3.1). Individually, each data source may have been designed for a specific, limited purpose. But when multiple sources are processed by techniques of modern statistical data mining, pattern recognition, and the combining of records from diverse sources by virtue of common identifying data, new meanings can be found. In particular, data fusion frequently results in the identification of individual people (that is, the association of events with unique personal identities), the creation of data-rich profiles of an individual, and the tracking of an individual's activities over days, months, or years.

By definition, the privacy challenges from data fusion do not lie in the individual data streams, each of whose collection, real-time processing, and retention may be wholly necessary and appropriate for its overt, immediate purpose. Rather, the privacy challenges are emergent properties of our increasing ability to bring into analytical juxtaposition large, diverse data sets and to process them with new kinds of mathematical algorithms.

3.1.2. Data from Sensors

Turn now to the second broad class of information emanations. One can say that information is "born analog" when it arises from the characteristics of the physical world. Such information does not become accessible electronically until it impinges on a "sensor," an engineered device that observes physical effects and converts them to digital form. The most common sensors are cameras, including video, which sense visible electromagnetic radiation; and microphones, which sense sound and vibration. There are many other kinds of sensors, however. Today, cell phones routinely contain not only cameras, microphones, and radios but also analog sensors for magnetic fields (3-D compass) and motion (acceleration). Other kinds of sensors include those for thermal infrared (IR) radiation; air quality, including the identification of chemical pollutants; barometric pressure (and altitude); low-level gamma radiation; and many other phenomena.

Examples of born-analog data providing personal information and in use today include:

- the voice and/or video content of a phone call — born analog but immediately converted to digital by the phone's microphone and camera
- personal health data such as heartbeat, respiration, and gait, as sensed by special-purpose devices (Fitbit has been a leading provider[61]) or cell-phone apps
- cameras/sensors in televisions and video games that interpret gestures by the user
- video from security surveillance cameras, mobile phones, or overhead drones
- imaging infrared video that can see in what people perceive as total darkness (and also see evanescent traces of past events, so-called heat scars)
- microphone networks in cities, used to detect and locate gunshots and for public safety
- cameras/microphones in classrooms and other meeting rooms
- ultrasonic motion detectors
- medical imaging, CT, and MRI scans, ultrasonic imaging

- opportunistically collected chemical or biological samples, notably trace DNA (today requiring slow, off-line analysis, but foreseeably more nimble)
- synthetic aperture radar (SAR), which can image through clouds and, under some conditions, see inside of non-metallic structures
- unintended radiofrequency emissions from electrical and electronic devices

When data are born analog, they are likely to contain more information than the minimum necessary for their immediate purpose, for several valid reasons. One is that the desired information ("signal") must be sensed in the presence of unwanted extraneous information ("noise"). The technologies typically work by sensing the environment ("signal plus noise") with high precision, so that mathematical techniques can then be applied that will separate the two even in the worst anticipated case when the signal is smallest or the noise is largest.

Another reason is technological convergence. For example, as the cameras in cell phones become smaller and cheaper, the use of identical components in other products becomes a favored design choice, even when full images are not needed. Where a big-screen television today has separate sensors for its IR remote control, room brightness, and motion detection (a feature that turns off the picture when no one is in the room), plus a true video camera in the add-on game console, tomorrow's model may integrate all of these functions in a single, cheap, high-resolution, IR-sensitive camera, a few millimeters in size.

In addition to the information available from digital and analog sources consciously intended to provide information about people, inadvertent disclosure abounds from the emerging "Internet of Things," an amalgamation of sensors whose primary purpose is enhanced by "smart" network-connected computational capabilities. Examples include "smart" thermostats that detect human presence and adjust air temperatures accordingly, "smart" automobile-ignition systems, and locking systems that are biometrically triggered.

The privacy challenges of born-analog data are somewhat different from those of born-digital data. Where over-collection (as was defined above) is an irrational design choice for the principled digital designer – and therefore an identifiable red flag for privacy issues – over-collection in the analog domain can be a robust and economical design choice. A consequence is that born-analog data will often contain information that was not originally expected. Unexpected information could in many cases lead to unanticipated beneficial

Big Data and Privacy: A Technological Perspective   125

products and services, but it could also give opportunities for unanticipated misuse.

As a concrete example, one might consider three key parameters of video imaging: resolution (how many pixels in the image), contrast ratio (how well can the image see into dark regions), and photometric precision (how accurate is the image in brightness and color). All three parameters have improved by orders of magnitude and are likely to keep improving. Today, with special cameras, one can image a cityscape from a high rooftop and see clearly into every facing house and apartment window within several miles.[62] Or, already mentioned, the ability exists to sense remotely the pulse of an individual, giving information on health status and emotional state.[63]

It is foreseeable, perhaps inevitable, that these capabilities will be present in every cell phone and security- surveillance camera, or every wearable computer device. (Imagine the process of negotiating the price for a car, or negotiating an international trade agreement, when every participant's Google Glass (or security camera or TV camera) is able to monitor and interpret the autonomic physiological state of every other participant, in real time.) It is unforeseeable what other unexpected information also lies in signals from the same sensors.

Once they enter the digital world, born-analog data can be fused and mined along with born-digital data. For example, facial-recognition algorithms, which might be error-prone in isolation, may yield nearly perfect identity tracking when they can be combined with born-digital data from cell phones (including unintended emanations), point-of-sale transactions, RFID tags, and so forth; and also with other born-analog data such as vehicle tracking (e.g., from overhead drones) and automated license-plate reading. Biometric data can provide identity information that enhances the profile of an individual even more, and data on behavior (as from social networks) are being used to analyze attitudes or emotions ("sentiment analysis," for individuals or groups[64]). In short, more and more information can be captured and put in a quantified format so it can be tabulated and analyzed.[65]

3.2. Big Data Analytics

Analytics is what makes big data come alive. Without analytics, big datasets could be stored, and they could be retrieved, wholly or selectively. But what comes out would be exactly what went in. Analytics, comprising a number of different computational technologies, is what fuels the big-data revolution.[66] Analytics is what creates the new value in big datasets, vastly more than the sum of the values of the parts.[67]

3.2.1. Data Mining

Data-mining, sometimes loosely equated to analytics but actually only a subset of it, refers to a computational process that discovers patterns in large data sets. It is a convergence of many fields of academic research in both applied mathematics and computer science, including statistics, databases, artificial intelligence, and machine learning. Like other technologies, advances in data mining have a research and development stage, in which new algorithms and computer programs are developed, and they have subsequent phases of commercialization and application.

Data mining algorithms can be trained to find patterns either by supervised learning, so-called because the algorithm is seeded with manually curated examples of the pattern to be recognized, or by unsupervised learning, where the algorithm tries to find related pieces of data without prior seeding. A recent success of unsupervised-learning algorithms was a program that, searching millions of images on the web, figured out on its own that "cat" was a much-posted category.[68]

The desired output of data mining can take several forms, each with its own specialized algorithms.[69]

- Classification algorithms attempt to assign objects or events to known categories. For example, a hospital might want to classify discharged patients as high, medium, or low risk for readmission.
- Clustering algorithms group objects or events into categories by similarity, as in the "cat" example above.
- Regression algorithms (also called numerical prediction algorithms) try to predict numerical quantities. For example, a bank may want to predict, from the details in a loan application, the probability of a default.
- Association techniques try to find relationships between items in their data set. Amazon's suggested products and Netflix's suggested movies are examples.
- Anomaly-detection algorithms look for untypical examples within a data set, for example, detecting fraudulent transactions on a credit-card account.
- Summarization techniques attempt to find and present salient features in data. Examples include both simple statistical summaries (e.g.,

average student test scores by school and teacher), and higher-level analysis (e.g., a list of key facts about an individual as gleaned from all web postings that mention her).

Data mining is sometimes confused with machine learning, the latter a broad subfield of computer science in academic and industrial research.[70] Data mining makes use of machine learning, as well as other disciplines, while machine learning has applications to fields other than data mining, for example, robotics.

There are limitations, both practical and theoretical, to what data mining can accomplish, as well as limits to how accurate it can be. It may reveal patterns and relationships, but it usually cannot tell the user the value or significance of these patterns. For example, supervised learning based on the characteristics of known terrorists might find similar persons, but they might or might not be terrorists; and it would miss different classes of terrorists who don't fit the profile.

Data mining can identify relationships between behaviors and/or variables, but these relationships do not always indicate causality. If people who live under high-voltage power lines have higher morbidity, it might mean that power lines are a hazard to public health; or it might mean that people who live under power lines tend to be poor and have inadequate access to health care. The policy implications are quite different. While so-called confounding variables (in this example, income) can be corrected for when they are known and understood, there is no sure way to know whether all of them have been identified. Imputing true causality in big data is a research field in its infancy.[71]

Many data analyses yield correlations that might or might not reflect causation. Some data analyses develop imperfect information, either because of limitations of the algorithms, or by the use of biased sampling. Indiscriminate use of these analyses may cause discrimination against individuals or a lack of fairness because of incorrect association with a particular group.[72] In using data analyses, particular care must be taken to protect the privacy of children and other protected groups.

Real-world data are incomplete and noisy. These data-quality issues lower the performance of data-mining algorithms and obscure outputs. When economics allow, careful screening and preparation of the input data can improve the quality of results, but this data preparation is often labor intensive and expensive. Users, especially in the commercial sector, must trade off cost and accuracy, sometimes with negative consequences for the individual represented in the data. Additionally, real-world data can contain extreme events or outliers. Outliers may be real events that, by chance, are overrepresented in the data; or they may be the result of data- entry or data-transmission errors. In both cases they can skew the model and degrade performance. The study of outliers is an important research area of statistics.

3.2.2. Data Fusion and Information Integration

Data fusion is the merging of multiple heterogeneous datasets into one homogeneous representation so that they can be better processed for data mining and management. Data fusion is used in a number of technical domains such as sensor networks, video/image processing, robotics and intelligent systems, and elsewhere.

Data integration is differentiated from data fusion in that integration more broadly combines data sets and retains the larger set of information. In data fusion, there is usually a reduction or replacement technique. Data fusion is facilitated by data interoperability, the ability for two systems to communicate and exchange data.

Data fusion and data integration are key techniques for business intelligence. Retailers are integrating their online, in-store, and catalog sales databases to create more complete pictures of their customers. Williams-Sonoma, for example, has integrated customer databases with information on 60 million households. Variables including household income, housing values, and number of children are tracked. It is claimed that targeted emails based on this information yield ten to 18 times the response rate of emails that are not targeted.[73] This is a simple illustration of how more information can lead to better inferences. Techniques that can help to preserve privacy are emerging.[74]

There is a great amount of interest today in multi-sensor data fusion.[75] The biggest technical challenges being tackled today, generally through development of new and better algorithms, relate to data precision/resolution, outliers and spurious data, conflicting data, modality (both heterogeneous and homogeneous data) and dimensionality, data correlation, data alignment, association within data, centralized vs. decentralized processing, operational timing, and the ability to handle dynamic vs. static phenomena. Privacy concerns may arise from sensor fidelity and precision as well as correlation of data from multiple sensors. A single sensor's output might not be sensitive, but the combination from two or more may raise privacy concerns.

3.2.3. Image and Speech Recognition

Image- and speech-recognition technologies are able to extract information, in some limited cases approaching human understanding, from massive corpuses of still images, videos, and recorded or broadcast speech.

Urban-scene extraction can be accomplished using a variety of data sources from photos and videos to ground based LiDAR (a remote-sensing technique using lasers).[76] In the government sector, city models are becoming vital for urban planning and visualization. They are equally important for a broad range of academic disciplines including history, archeology, geography, and computer-graphics research. Digital city models are also central to popular consumer mapping and visualization applications such as Google Earth and Bing Maps, as well as GPS- enabled navigation systems.[77] Scene extraction is an example of the inadvertent capture of personal information and can be used for data fusion that reveals personal information.

Facial-recognition technologies are beginning to be practical in commercial and law-enforcement applications.[78] They are able to acquire, normalize, and recognize moving faces in dynamic scenes. Real-time video surveillance with single-camera systems (and some with multi-camera systems, which can both recognize objects and analyze activity) has a wide variety of applications in both public and private environments, such as homeland security, crime prevention, traffic control, accident prediction and detection, and monitoring patients, the elderly, and children at home.[79] Depending on the application, use of video surveillance is at varying levels of deployment.[80]

Additional capabilities of image recognition include

- Video summarization and scene-change detection (that is, picking the small number of images that summarize a period of time)
- Precise geolocation in imagery from satellites or drones
- Image-based biometrics
- Human-in-the-loop surveillance systems
- Re-identification of persons and vehicles, that is, tracking the same person or vehicle as it moves from sensor to sensor
- Human-activity recognition of various kinds
- Semantic summarization (that is, converting pictures into text summaries)

Although systems are expected to become able to track objects across camera views and detect unusual activities in a large area by combining information from multiple sources, re-identification of objects remains hard to do (a challenge for inter-camera tracking), as is video surveillance in crowded environments.

Although the data they use are often captured in public areas, scene-extraction technologies like Google Street View have triggered privacy concerns. Photos captured for use in Street View may contain sensitive information about people who are unaware they are being observed and photographed.[81]

Social-media data can be used as an input source for scene extraction techniques. When these data are posted, however, users are unlikely to know that their data would be used in these aggregated ways and that their social media information (although public) might appear synthesized in new forms.[82]

Automated speech recognition has existed since at least the 1950s,[83] but recent developments over the last 10 years have allowed for novel new capabilities. Spoken text (e.g., news broadcasters reading part of a document) can today be recognized with accuracy higher than 95 percent using state-of-the-art techniques. Spontaneous speech is much harder to recognize accurately. In recent years there has been a dramatic increase in the corpuses of spontaneous speech data available to researchers, which has allowed for improved accuracy.

Over the next few years speech-recognition interfaces will be in many more places. For example, multiple companies are exploring speech recognition to control televisions and cars, to find a show on TV, or to schedule a DVR recording. Researchers at Nuance say they are actively planning how speech technology would have to be designed to be available on wearable computers.[84] Google has already implemented some of this basic functionality in its Google Glass product, and Microsoft's Xbox One system already integrates machine vision and multi-microphone audio input for controlling system functions.

3.2.4. Social-Network Analysis

Social-network analysis refers to the extraction of information from a variety of interconnecting units under the assumption that their relationships are important and that the units do not behave autonomously.[85] Social networks often emerge in an online context. The most obvious examples are dedicated online social media platforms, such as Facebook, LinkedIn and Twitter, which provide new access to social interaction by allowing users to connect directly with each other over the Internet to communicate and share information. Offline human social networks may also leave analyzable digital traces, such as in phone-call metadata records that record which phones have exchanged calls or texts, and for how long. Analysis of social networks is increasingly enabled by the rising collection of digital data that links people together, especially when it is correlated to other data or metadata about the individual.[86] Tools for such analysis are being developed and made available,[87] motivated in part by the growing amount of social network content accessible through open application-programming interfaces to online social-media platforms. This sort of analysis is an active arena for research.

Social-network analysis complements analysis of conventional databases, and some of the techniques used (e.g., clustering in association networks) can be used in either context. Social-network analysis can be more powerful because of the easy association of diverse kinds of information (i.e., considerable data fusion is possible). It lends itself to visualization of the results, which aids in interpreting the results of the analysis. It can be used to learn about people through their association with others, in a context of people's tendency to associate with others who are have some similarities to themselves.[88]

Social-network analysis is yielding results that may surprise people. In particular, unique identification of an individual is easier than from database analysis alone. Moreover, it is achieved through more diverse kinds of data than many people may understand, contributing to the erosion of anonymity.[89] The structure of an individual's network is unique and itself serves as an identifier; co-occurrence in time and space is a significant means of identification; and, as discussed elsewhere in this report, different kinds of data can be combined to foster identification.[90]

Social-network analysis is used in criminal forensic investigations to understand the links, means, and motives of those who may have committed crimes. In particular, social-network analysis has been used to better understand covert terrorist networks, whose dynamics may be different from those of overt networks.[91]

In the realm of commerce, it is well-understood that what a person's friends like or buy can influence what he or she might buy. For example, in 2010, it was reported that having one iPhone-owning friend makes a person three times more likely to own an iPhone than otherwise. A person with two iPhone-owning friends was five times more likely to have one.[92] Such correlations emerge in social-network analysis and can be used to help predict product trends, tailor marketing campaigns towards products an individual may be more likely to want, and target customers (said to have higher "network value") with a central role (and a large amount of influence) in a social network.[93]

Because disease is commonly spread via direct contact between individuals (humans or animals), understanding social networks through whatever proxies are available can suggest possible direct contacts and thereby assist in monitoring and stemming the outbreak of disease.

A recent study by researchers at Facebook analyzed the relationship between geographic location of individual users and that of their friends. From this analysis, they were able to create an algorithm to predict the location of an individual user based upon the locations of a small number of friends in their network, with higher accuracy than simply looking at the user's IP address.[94]

There are many commercial "social listening" services, such as Radian6/Salesforce Cloud, Collective Intellect, Lithium, and others, that mine data from social-networking feeds for use in business intelligence.[95] Coupled with social-network analysis, this information can be used to evaluate changing influences and the spread of trends between individuals and communities to inform marketing strategies.

3.3. The Infrastructure behind Big Data

Big-data analytics requires not just algorithms and data, but also physical platforms where the data are stored and analyzed. The related security services used for personal data (see Sections 4.1 and 4.2) are also an essential component of the infrastructure. Once available only to large organizations, this class of infrastructure is now available through "the cloud" to small businesses and to individuals. To the extent that the software infrastructure is widely shared, privacy-preserving infrastructure services can also be more readily used.

3.3.1. Data Centers

One way to think about big-data platforms is in physical units of "data centers." In recent years, data centers have become almost standard commodities. A typical data center is a large, warehouse-like building on a concrete slab the size of a few football fields. It is located with good access to cheap electric power and to a fiber-optic, Internet-backbone connection, usually in a rural or isolated area. The typical center consumes 20-40 megawatts of power (the equivalent of a city with 20,000-40,000 residents) and today houses some tens of thousands of servers and hard-disk drives, totaling some tens of petabytes.[96] Worldwide, there are roughly 6000 data centers of this scale, about half in the United States.[97]

Data centers are the physical locus of big data in all its forms. Large data collections are often replicated in multiple data centers to improve both performance and robustness. There is a growing marketplace in selling data-center services.

Specialized software technology allows the data in multiple data centers (and spread across tens of thousands of processors and hard-disk drives) to cooperate in performing the tasks of data analytics, thereby providing both scaling and better performance. For example, MapReduce (originally a proprietary technology of Google, but now a term used generically) is a programming model for parallel operations across a practically unlimited number of processors; Hadoop is a popular open-source programming platform and program library based on the same ideas; NoSQL (the name derived from "*not* Structured Query Language") is a set of database technologies that relaxes many of the restrictions of traditional, "relational" databases and allows for better scalability across the many processors in one or more data centers. Contemporary research is aimed at the next generation beyond Hadoop. One path is represented by Accumulo, initiated by the National Security Agency and transitioned to the open-source Apache community.[98] Another is the Berkeley Data Analytics Stack, an open-source platform that outperforms Hadoop by a factor of 100 for memory-intensive data analytics and is being used by such companies as Foursquare, Conviva, Klout, Quantifind, Yahoo, and Amazon Web Services.[99] Sometimes termed "NoHadoop" (to parallel the movement from SQL to NoSQL), technologies that fit this trend include Google's Dremel, MPI (typically used in supercomputing), Pregel (for graphs), and Cloudscale (for real-time analytics).

3.3.2. The Cloud

The "cloud" is not just the world inventory of data centers (although much of the public may think of it as such). Rather, one way of understanding the cloud is as a set of platforms and services *made possible* by the physical commoditization of data centers. When one says that data are "in the cloud," one refers not just to the physical hard-disk drives that exist (somewhere!) with the data, but also to the complex infrastructure of application programs, middleware, networking protocols, and (not least) business models that allow that data to be ingested, accessed, and utilized, all with costs that are competitively allocated. The commercial entities that, in aggregate, provision the cloud exist in an ecosystem that has many hierarchical levels and many different coexisting models of value added. There may be several handoffs of responsibility between the end user and the physical data center.

Today's cloud providers offer some security benefits (and through that, privacy benefits) as compared to yesterday's conventional corporate data centers or small-business computers.[100] These services may include better physical protection and monitoring, as well as centralized support staffing, training, and oversight. Cloud services also pose new challenges for security, a subject of current research. Both benefits and risks come from the centralization of resources: More data are held by a given entity (albeit distributed across multiple servers or sites), and a cloud provider can perform better than separately held data centers by applying high standards to recruiting and managing people and systems.

Usage of the cloud and individual interactions with it (whether witting or not) are expected to increase dramatically in coming years. The rise of both mobile apps,[101] reinforcing the use of cell phones and tablets as platforms, and broadly distributed sensors is associated with the growing use of cloud systems for storing, processing, and otherwise acting on information contributed by dispersed devices. Although progress in the mobile environment improves the usability of mobile cloud applications, it may be detrimental to privacy to the extent that it more effectively hides information exchange from the user. As more core mobile functionality is transitioned to the cloud, larger amounts of information will be exchanged, and users may be surprised by the nature of the information that no longer remains localized to their cell phone. For example, cloud-based screen rendering (or "virtualized screens") for cell phones would mean that the images shown on a cell-phone screen will actually be calculated on the cloud and transmitted to the mobile device. This means all the images on the screen of the mobile device can be accessed and manipulated from the cloud.

Cloud architectures are also being used increasingly to support big-data analytics, both by large enterprises (e.g., Google, Amazon, eBay) and by small entities or individuals who make ad hoc or routine use of public cloud platforms (e.g., Amazon Web Services, Google Cloud Platform, Microsoft Azure) in lieu of acquiring their own infrastructure. Social-media services such as Facebook and Twitter are deployed and analyzed by their providers using cloud systems. These uses represent a kind of democratization of analytics, with the potential to facilitate new businesses and more. Prospects for the future include exploration of options for federating or interconnecting cloud applications and for reducing some of the heterogeneity in application-programming interfaces for cloud applications.[102]

4. TECHNOLOGIES AND STRATEGIES FOR PRIVACY PROTECTION

Data come into existence, are collected, and are possibly processed immediately (including adding "metadata"), possibly communicated, possibly stored (locally, remotely, or both), possibly copied, possibly analyzed, possibly communicated to users, possibly archived, possibly discarded. Technology at any of these stages can affect privacy positively or negatively.

This section focuses on the positive and assesses some of the key technologies that can be used in service of the protection of privacy. It seeks to clarify the important distinctions between privacy and (cyber-)security, as well as the vital, but yet limited, role that encryption technology can play. Some older techniques, such as anonymization, while valuable in the past, are seen as having only limited future potential. Newer technologies, some entering the marketplace and some requiring further research, are summarized.

4.1. The Relationship between Cybersecurity and Privacy

Cybersecurity is a discipline, or set of technologies, that seeks to enforce policies relating to several different aspects of computer use and electronic communication.[103] A typical list of such aspects would be

- identity and authentication: Are you who you say you are?
- authorization: What are you allowed to do?
- availability: Can attackers interfere with authorized functions?
- confidentiality: Can data or communications be (passively) copied by someone not authorized to do so?
- integrity: Can data or communications be (actively) changed or manipulated by someone not authorized?
- non-repudiation, auditability: Can actions (payments may provide the best example) later be shown to have occurred?

Good cybersecurity enforces policies that are precise and unambiguous. Indeed, such clarity of policy, expressible in mathematical terms, is a necessary prerequisite for the Holy Grail of cybersecurity," provably secure" systems. At present, provable security exists only in very limited domains, for example, for certain functions on some kinds of computer chips. It is a goal of cybersecurity research to extend the scope of provably secure systems to larger and larger domains. Meanwhile, practical cybersecurity draws on the emerging principles of such research, but it is guided even more by practical lessons learned from known failures of cybersecurity. The realistic goal is that the practice of cybersecurity should be continuously improving so as to be, in most places and at most of the time, ahead of the evolving threat.

Poor cybersecurity is clearly a threat to privacy. Privacy can be breached by failure to enforce confidentiality of data, by failure of identity and authentication processes, or by more complex scenarios such as those compromising availability.

Security and privacy share a focus on malice. The security of data can be compromised by inadvertence or accident, but it can also be compromised because some party acted knowingly to achieve the compromise – in the language of security, committed an attack. Substituting the words "breach" or "invasion" for "compromise" or "attack," the same concepts apply to privacy.

Even if there were perfect cybersecurity, however, privacy would remain at risk. Violations of privacy are possible even when there is no failure in computer security. If an authorized individual chooses to misuse (e.g., disclose) data, what is violated is privacy policy, not security policy. Or, as we have discussed (see Section 3.1.1), privacy may be violated by the fusion of data – even if performed by authorized individuals on secure computer systems.[104]

Privacy is different from security in other respects. For one thing, it is harder to codify privacy policies precisely. Arguably this is because the presuppositions and preferences of human beings have greater diversity than the useful scope of assertions about computer security. Indeed, how to codify human privacy preferences is an important, nascent area of research.[105]

When people provide assurance (at some level) that a computer system is secure, they are saying something about applications that are not yet invented: They are asserting that technological design features already in the machine today will prevent such application programs from violating pertinent security policies in that machine, even tomorrow.[106] Assurances about privacy are much more precarious. Since not-yet-invented applications will have access to not-yet-imagined new sources of data, as well as to not-yet-discovered powerful algorithms, it much harder to provide, today, technological safeguards against a new route to violation of privacy tomorrow. Security deals with tomorrow's threats against today's platforms. That is hard enough. But privacy deals with tomorrow's threats against *tomorrow's* platforms, since those "platforms" comprise not just hardware and software, but also new kinds of data and new algorithms.

Computer scientists often work from the basis of a formal policy for security, just as engineers aim to describe something explicitly so that they can design specific ways to deal with it by purely technical means. As more computer scientists begin to think about privacy, there is increasing attention to formal articulation of privacy policy.[107] To caricature, you have to know what you are doing to know whether what you are doing is doing the right thing.[108] Research addressing the challenges of aligning regulations and policies with software specifications includes formal languages to express policies and system requirements; tools to reason about conflicts, inconsistencies, and ambiguities within and among policies and software specifications; methods to enable requirements engineers, business analysts, and software developers to analyze and refine policy into measurable system specifications that can be monitored over time; formalizing and enforcing privacy through auditing and accountability systems; privacy compliance in big-data systems; and formalizing and enforcing purpose restrictions.

4.2. Cryptography and Encryption

Cryptography comprises a set of algorithms and system-design principles, some well-developed and others nascent, for protecting data. Cryptography is a field of knowledge whose products are encryption technology. With well-designed protocols, encryption technology is an inhibitor to compromising privacy, but it is not a "silver bullet."[109]

4.2.1. Well-Established Encryption Technology

Using cryptography, readable data of any kind, termed plaintext, are transformed into what are, for all intents and purposes, incomprehensible strings of provably random bits, so-called cryptotext. Cryptotext requires no security protection of any kind. It can be stored in the cloud or sent anywhere that is convenient. It can be sent promiscuously to both the NSA and Russian FSB. If they have only cryptotext – and if it was properly generated in a precise mathematical sense – it is useless to them. They can neither read the data nor compute with it. What is needed to decrypt, to turn cryptotext back into the original plaintext, is a "key," which is in practice a string of bits that is supposed to be known to (or computable by) only authorized users. Only with the key can encrypted data be used, i.e., their value read.

In the context of protecting privacy, it is primarily not the cryptography that is of concern.[110] Rather, compromises of data will occur in one of two main ways:

- Data can be stolen, or mistakenly shared, before they have been encrypted or after they have been decrypted. Many attacks on supposedly encrypted data are actually attacks on machines that contain – however briefly – unencrypted plaintext. For example, in Target's 2013 breach of one hundred million debit card number and personal-identification numbers (PINs), the PINs were present in unencrypted form only ephemerally. They were stolen nonetheless.[111]
- Keys must be authorized, generated, distributed, and used. At every stage of a key's life, it is potentially open to compromise or misuse that can ultimately compromise the data that the key was intended to protect. No system based on encryption is secure, of course, if persons with access to private keys can be coerced into sharing them.

Until the 1970s, keys were distributed physically, on paper or computer media, protected by registered mail, armed guards, or anything in between. The invention of "public-key cryptography"[112] changed everything. Public-key cryptography, as the name implies, allows individuals to broadcast publicly their personal key. But this public key is only an encryption key, useful for turning plaintext into cryptotext that is meaningless to others. Its corresponding "private key," used to transform cryptotext to plaintext, is still kept secret by the recipient. Public-key cryptography thus turns the problem of key distribution into a problem of identity determination. Alice's messages (encrypted data transmissions) to Bob are completely protected by Bob's public key – but only if Alice is certain that it is really *Bob's* public key that she is using, and not the public key of someone merely masquerading as Bob.

Luckily, public-key cryptography also provides some techniques for helping to establish identity, namely the electronic "signing" of messages to document their authenticity. Electronic signatures, in turn, enable messages of the form "I, a person of authority known as X, certify that the following is really the public key of subordinate person Y. (Signed) X." Messages like this are termed certificates. Certificates can be cascaded, with A certifying the identity of B, who certifies C, and so on. Certificates essentially transform the identity problem from one of validating the identity of millions of possible Y's to validating the identity of much smaller number of top-level certificate authorities (CAs). Yet it is a matter of concern that more than 100 top-level

CAs are widely recognized (e.g., accepted by most all web browsers), because there may be several intermediate steps in the hierarchy of certificates from a CA to a user, and at every step a private key must be protected by some signer on some computer. The compromise of this private key potentially compromises the privacy of all users lower down the chain – because forged certificates of identity can now be created. Such exploits have been seen. For example, the 2011 apparent theft of a Dutch CA's private key compromised the privacy of potentially all government records in the Netherlands.[113,114]

Many major companies have recently introduced or strengthened their use of encryption to transmit data.[115] Some are now using "(perfect) forward secrecy," a variant of public-key cryptography that ensures that the compromise of an individual's private key can compromise only messages that he receives subsequently, while the confidentiality of past conversations is maintained, even if their cryptotext was previously recorded by the same eavesdropper now in possession of the purloined private key.[116]

4.2.2. Encryption Frontiers

The technologies thus far mentioned enable the protection of data both in storage and in transit, allowing those data to be fully decrypted by users who either (i) have the right key already (as might be the case for persons storing data for their own later use), or (ii) are authorized by the data owner and have identities certified by a CA that is itself trusted by the data owner. A frontier of cryptography research, with some inventions now starting to make it into practice, is how to create different kinds of keys, ones which give only limited access of various kinds, or which allow messages to be sent to classes of individuals without knowing in advance exactly who they may be.

For example, "identity-based encryption" and "attribute-based encryption" are ways of sending a message, or protecting a file of data, for the exclusive use of "a person named Ramona Q. Doe who was born on May 23, 1980," or for "anyone with the job title ombudsman, ombudsperson, or consumer advocate." These techniques require a trusted third party (essentially a certificate authority), but the messages themselves do not need to pass through the hands of that third party. These tools are in early stages of adoption.

"Zero-knowledge" systems allow encrypted data to be queried for certain higher-level abstractions without revealing the low-level data. For example, a website operator could verify that a user is over age 21 without learning the user's actual birthdate. What is remarkable is that this can be done in a way that proves mathematically that the user is not lying about his age: The operator learns with mathematical certainty that a certificate (signed by some

CA of course!) attests to the user's birthdate, without ever actually seeing that certificate. Zero-knowledge systems are just beginning to be commercialized in simple cases. They are not foreseeably extendable to complex and unstructured situations, such as what might be needed for the research mining of health-record data from non-consenting patients.

In some simpler domains, for example location privacy, practical cryptographic protection is closer to reality. The typical case might be that a group of friends want to know when they are close to one another, but without sharing their actual locations with any third party. Applications like this are, of course, much simpler if there is a trusted third party, as is *de facto* the case for most such commercial applications today.

Homomorphic encryption is a research area that goes beyond the mere querying of encrypted databases to actual computations (e.g., the collection of statistics) using encrypted data without ever decrypting it. These techniques are far from being practical, and they are unlikely to provide policy options on the timescale relevant to this report.

In secure multi-party computation, which is related to homomorphic encryption and is of particular interest in the financial sector, computation may be done on distributed data stores that are encrypted. Although individual data are kept private using "collusion-robust" encryption algorithms, data can be used to calculate general statistics. Parties that each know some private data use a protocol that generates useful results based on both information they know and information they do not know, without revealing to them data they do not already know.

Differential privacy, a comparatively new development related to but different from encryption, aims to maximize the accuracy of database queries or computations while minimizing the identifiability of individuals with records in the database, typically via obfuscation of query results (for example, by the addition of spurious information or "noise").[117] As with other obfuscation approaches, there is a tradeoff between data anonymity and the accuracy and utility of the query outputs. These ideas are far from practical application, except insofar as they may enable the risks of allowing any queries at all to be better assessed.

4.3. Notice and Consent

Notice and consent is, today, the most widely used strategy for protecting consumer privacy. When the user downloads a new app to his or her mobile

Big Data and Privacy: A Technological Perspective 141

device, or when he or she creates an account for a web service, a notice is displayed, to which the user must positively indicate consent before using the app or service. In some fantasy world, users actually read these notices, understand their legal implications (consulting their attorneys if necessary), negotiate with other providers of similar services to get better privacy treatment, and only then click to indicate their consent. Reality is different.[118]

Notice and consent fundamentally places the burden of privacy protection on the individual – exactly the opposite of what is usually meant by a "right." Worse yet, if it is hidden in such a notice that the provider has the right to share personal data, the user normally does not get any notice from the next company, much less the opportunity to consent, even though use of the data may be different.

Furthermore, if the provider changes its privacy notice for the worse, the user is typically not notified in a useful way.

As a useful policy tool, notice and consent is defeated by exactly the positive benefits that big data enables: new, non-obvious, unexpectedly powerful uses of data. It is simply too complicated for the individual to make fine-grained choices for every new situation or app. Nevertheless, since notice and consent is so deeply rooted in current practice, some exploration of how its usefulness might be extended seems warranted.

One way to view the problem with notice and consent is that it creates a non-level playing field in the implicit privacy negotiation between provider and user.

The provider offers a complex take-it-or-leave-it set of terms, backed by a lot of legal firepower, while the user, in practice, allocates only a few seconds of mental effort to evaluating the offer, since acceptance is needed to complete the transaction that was the user's purpose, and since the terms are typically difficult to comprehend quickly. This is a kind of market failure. In other contexts, market failures like this can be mitigated by the intervention of third parties who are able to represent significant numbers of users and negotiate on their behalf. Section 4.5.1 below suggests how such intervention might be accomplished.

4.4. Other Strategies and Techniques

4.4.1. Anonymization or De-Identification

Long used in health-care research and other research areas involving human subjects, anonymization (also termed de-identification) applies when the data, standing alone and without an association to a specific person, do not violate privacy norms. For example, you may not mind if your medical record is used in research as long as you are identified only as Patient X and your actual name and patient identifier are stripped from that record.

Anonymization of a data record might seem easy to implement. Unfortunately, it is increasingly easy to defeat anonymization by the very techniques that are being developed for many legitimate applications of big data. In general, as the size and diversity of available data grows, the likelihood of being able to re-identify individuals (that is, re-associate their records with their names) grows substantially.[119]

One compelling example comes from Sweeney, Abu, and Winn.[120] They showed in a recent paper that, by fusing public, Personal Genome Project profiles containing zip code, birthdate, and gender with public voter rolls, and mining for names hidden in attached documents, 84-97 percent of the profiles for which names were provided were correctly identified.

Anonymization remains somewhat useful as an added safeguard, but it is not robust against near-term future re-identification methods. PCAST does not see it as being a useful basis for policy. Unfortunately, anonymization is already rooted in the law, sometimes giving a false expectation of privacy where data lacking certain identifiers are deemed not to be personally identifiable information and therefore not covered by such laws as the Family Educational Rights and Privacy Act (FERPA).

4.4.2. Deletion and Non-Retention

It is an evident good business practice that data of all kinds should be deleted when they are no longer of value. Indeed, well-run companies often mandate the destruction of some kinds of records (both paper and electronic) after specified periods of time, often because they see little benefit in keeping the records as well as potential cost in producing them. For example, employee emails, which may be subject to legal process by (e.g.) divorce lawyers, are often seen as having negative retention value.

Counter to this practice is the new observation that big data is frequently able to find economic or social value in masses of data that were otherwise considered to be worthless. As the physical cost of retention continues to decrease exponentially with time (especially in the cloud), there will be a tendency in both government and the private sector to hold more data for longer – with obvious privacy implications. Archival data may also be important to future historians, or for later longitudinal analysis by academic researchers.

Only policy interventions will counter this trend. Government can mandate retention policies for itself. To affect the private sector, government may mandate policies where it has regulatory authorities (as for consumer protection, for example). But it can also encourage the development of stricter liability standards for companies whose data, including archived data, cause harm to individuals. A rational response by the private sector would then be to hold fewer data or to protect their use.

The above holds true for privacy-sensitive data about individuals that are held overtly – that is, the holder knows that he has the data and to whom they relate. As was discussed in Section 3.1.2, however, sources of data increasingly contain latent information about individuals, information that becomes known only if the holder expends analytic resources (beyond what may be economically feasible), or that may become knowable only in the future with the development of new data-mining algorithms. In such cases it is practically impossible for the data holder even to surface "all the data about an individual," much less delete those data on any specified schedule.

The concepts of ephemerality (keeping data only on-the-fly or for a brief period), and transparency (enabling the individual to know what data about him or her are held) are closely related, and with the same practical limitations. While data that are only streamed, and not archived, may have lower risk of future use, there is no guarantee that a violator will play by the supposed rules, as in Target's loss of 100 million debit card PINs, each present only ephemerally (see Section 4.2.1).

Today, given the distributed and redundant nature of data storage, it is not even clear that data *can* be destroyed with any useful degree of assurance. Although research on data destruction is ongoing, it is a fundamental fact that at the moment that data are displayed (in "analog") to a user's eyeballs or ears, they can also be copied ("re-digitized") without any technical protections. The same holds if data are ever made available in unencrypted form to a rogue computer program, one designed to circumvent technical safeguards. Some misinformed public discussion notwithstanding, there is no such thing as automatically self-deleting data, other than in a fully controlled and rule-abiding environment.

As a current example, SnapChat provides the service of delivering ephemeral snapshots (images), visible for only a few seconds, to a designated recipient's mobile device. SnapChat promises to delete past-date snaps from their servers, but it is only a promise. And, they are careful *not* to promise that the intended recipient may not contrive to make an uncontrolled and non-expiring copy. Indeed, the success of SnapChat incentivizes the development of just such copying applications.[121]

From a policymaking perspective, the only viable assumption today, and for the foreseeable future, is that data, once created, are permanent. While their *use* may be regulated, their continued *existence* is best considered conservatively as unalterable fact.

4.5. Robust Technologies Going Forward

4.5.1. A Successor to Notice and Consent

The purpose of notice and consent is that the user assents to the collection and use of personal data for a stated purpose that is acceptable to that individual. Given the large number of programs and Internet-available devices, both visible and not, that collect and use personal data, this framework is increasingly unworkable and ineffective. PCAST believes that the responsibility for using personal data in accordance with the user's preferences should rest with the provider, possibly assisted by a mutually accepted intermediary, rather than with the user.

How might that be accomplished? Individuals might be encouraged to associate themselves with one of a standard set of privacy preference profiles (that is, settings or choices) voluntarily offered by third parties. For example, Jane might choose to associate with a profile offered by the American Civil Liberties Union that gives particular weight to individual rights, while John might associate with one offered by *Consumer Reports* that gives weight to economic value for the consumer. Large app stores (such as Apple App Store, Google Play, Microsoft Store) for whom reputational value is important, or large commercial sectors such as finance, might choose to offer competing privacy-preference profiles.

In the first instance, an organization offering profiles would vet new apps as acceptable or not acceptable within each of their profiles. Basically, they would do the close reading of the provider's notice that the user should, but does not, do. This is not as onerous as it may sound: While there are millions of apps, the most popular downloads are relatively few and are concentrated in a relatively small number of portals. The "long tail" of apps with few customers each might initially be left as "unrated."

Simply by vetting apps, the third-party organizations would automatically create a marketplace for the negotiation of community standards for privacy. To attract market share, providers (especially smaller ones) could seek to

Big Data and Privacy: A Technological Perspective 145

qualify their offerings in as many privacy-preference profiles, offered by as many different third parties, as they deem feasible. The Federal government (e.g., through the National Institute of Standards and Technology) could encourage the development of standard, machine-readable interfaces for the communication of privacy implications and settings between providers and assessors.

Although human professionals could do the vetting today using policies expressed in natural language, it would be desirable in the future to automate that process. To do that, it would be necessary to have formalisms to specify privacy policies and tools to analyze software to determine conformance to those policies. But that is only part of the challenge. A greater challenge is to make sure the policy language is sufficiently expressive, the policies are sufficiently rich, and conformance tests are sufficiently powerful. Those requirements lead to a consideration of context and use.

4.5.2. Context and Use

The previous discussion, particularly that of Sections 3.1 and 3.2, illustrates PCAST's belief that a focus on the collection, storage, and retention of electronic personal data will not provide a technologically robust foundation on which to base future policy. Among the many authors that have touched on these issues, Kagan and Abelson explain why access control does not suffice to protect privacy.[122] Mundie gives a cogent and more complete explanation of this issue and advocates that privacy protection is better served by controlling the use of personal data, broadly construed, including metadata and data derived from analytics than by controlling collection.[123] In a complementary vein, Nissenbaum explains that both the context of usage and the prevailing social norms contribute to acceptable use.[124]

To implement in a meaningful way the application of privacy policies to the use of personal data for a particular purpose (i.e., in context), those policies need to be associated both with data and with the code that operates on the data. For example, it must be possible to ensure that only apps with particular properties can be applied to certain data. The policies might be expressed in what computer scientists call natural language (plain English or the equivalent) and the association done by the user, or the policies might be stated formally and their association and enforcement done automatically. In either case, there must also be policies associated with the outputs of the computation, since they are data as well. The privacy policies of the output data must be computed from the policies associated with the inputs, the policies associated with the code, and the intended use of the outputs (i.e., the context). These privacy properties are a kind of metadata. To achieve a reasonable level of reliability, their implementation must be tamper-proof and "sticky" when data are copied.

There has been considerable research in areas that would contribute to such a capability, some of which is beginning to be commercialized. There is a history of using metadata ("tags" or "attributes") in database systems to control use. While the formalization of privacy policies and their synthesis is a research topic,[125] manual interpretation of such policies and the human determination of usage tags can be found in recent products. Identity management systems (to authenticate users and their roles, i.e., their context) are also evident both in research[126] and in practice.[127]

Commercial privacy systems for implementing use control exist today under the name of Trusted Data Format (TDF) implementations, developed principally for the United States intelligence community.[128] TDF operates at the file level. The systems are primarily being implemented on a custom basis by large consulting firms, often assembled from open-source software components. Customers today are primarily government agencies, such as Federal intelligence agencies or local-government criminal intelligence units, or large commercial companies in vertically integrated industries like financial services and pharmaceutical companies looking to improve their accountability and auditing capabilities. Consulting services that have expertise in building such systems include, for example, Booz Allen, Ernst & Young, IBM, Northrop Grumman, and Lockheed; product-based companies like Palantir and new startups pioneering internal usage auditing, policy analytics, and policy reasoning engines have such expertise, as well. With sufficient market demand, more widespread market penetration could happen in the next five years. Market penetration would be further accelerated if the leading cloud-platform providers like Amazon, Google, and Microsoft implemented usage-controlled system technologies in their offerings. Wider-scale use through the government would help motivate the creation of off-the-shelf standard software.

4.5.3. Enforcement and Deterrence

Privacy policies and the control of use in context are only effective to the extent that they are realized and enforced. Technical measures that increase the probability that a violator is caught can be effective only when there are regulations and laws with civil or criminal penalties to deter the violators. Then there is both deterrence of harmful actions and incentive to deploy privacy-protecting technologies.

Big Data and Privacy: A Technological Perspective 147

It is today straightforward technically to associate metadata with data, with varying degrees of granularity ranging from an individual datum, to a record, to an entire collection. These metadata can record a wealth of auditable information, for example, provenance, detailed access and use policies, authorizations, logs of actual access and use, and destruction dates. Extending such metadata to derived or shared data (secondary use) together with privacy-aware logging can facilitate auditing. Although the state of the art is still somewhat ad hoc, and auditing is often not automated, so-called accountable systems are beginning to be deployed (Section 4.5.2). The ability to detect violations of privacy policies, particularly if the auditing is automated and continuous, can be used both to deter privacy violations and to ensure that violators are punished.

In the next five years, with regulation or market-driven encouragement, the large cloud-based infrastructure systems (e.g., Google, Amazon, Microsoft, Rackspace) could, as one example, incorporate the data-provenance and usage-compliance aspects of accountable systems into their cloud application-programming interfaces (APIs) and additionally provide APIs for policy awareness. These capabilities could then readily be included in open-source-based systems like Open Stack (associated with Rackspace)[129] and other provider platforms. Applications intended to run on such cloud-based systems could be built with privacy concepts "baked into them," even when they are developed by small enterprises or individual developers.

4.5.4. Operationalizing the Consumer Privacy Bill of Rights

In February 2012, the Administration issued a report setting forth a Consumer Privacy Bill of Rights (CPBR). The CPBR addresses commercial (not public sector) uses of personal data and is a strong statement of American privacy values.

For purposes of this discussion, the principles embodied in CPBR can be divided into two categories. First, there are obligations for data holders, analyzers, or commercial users. These are passive from the consumer's standpoint – the obligations should be met whether or not the consumer knows, cares, or acts. Second, and different, there are consumer empowerments, things that the consumer should be empowered to initiate actively. It is useful here to rearrange the CPBR's principles by category.

In the category of obligations are these elements:

- Respect for Context: Consumers have a right to expect that companies will collect, use, and disclose personal data in ways that are consistent with the context in which consumers provide the data.
- Focused Collection: Consumers have a right to reasonable limits on the personal data that companies collect and retain.
- Security: Consumers have a right to secure and responsible handling of personal data.
- Accountability: Consumers have a right to have personal data handled by companies with appropriate measures in place to assure they adhere to the Consumer Privacy Bill of Rights.

In the category of consumer empowerments are these elements:

- Individual Control: Consumers have a right to exercise control over what personal data companies collect from them and how they use it.
- Transparency: Consumers have a right to easily understandable and accessible information about privacy and security practices.
- Access and Accuracy: Consumers have a right to access and correct personal data in usable formats, in a manner that is appropriate to the sensitivity of the data and the risk of adverse consequences to consumers if the data are inaccurate.

PCAST endorses as sound the principles underlying CPBR. Because of the rapidly changing technologies associated with big data, however, effective operationalization of CPBR is at risk. Up to now, debate over how to operationalize CPBR has focused on the collection, storage, and retention of data, with an emphasis on the "small-data" contexts that motivated CPBR development. But, as discussed at multiple places in this report (e.g., Sections 3.1.2, 4.4 and 4.5.2), PCAST believes that such a focus will not provide a technologically robust foundation on which to base future policy that also applies to big data. Further, the increasing complexity of applications and uses of data undermines even a simple concept like "notice and consent."

PCAST believes that the principles of CPBR can readily be adapted to a more robust regime based on recognizing and controlling harmful uses of the data. Some specific suggestions follow.

Turn first to the rights classified above as obligations on the data holder.

The principle of Respect for Context needs augmentation. As this report has repeatedly discussed, there are instances in which personal data are not provided by the customer. Such data may emerge as a product of analysis well after the data were collected and after they may have passed through several hands. While the intent of the right is appropriate, namely that data be used for legitimate purposes that do not produce certain adverse consequences or harms to individuals, the CPBR's articulation in which "consumers provide the data" is too limited. This right needs to state in some way that data about an individual – however acquired – not be used so as to cause certain adverse consequences or harms to that individual. (See Section 1.4 for a possible list of adverse consequences and harms that might be subject to some regulation.)

As initially conceived, the right to Focused Collection was to be achieved by techniques like de-identification and data deletion. As discussed in Section 4.4.1, however, de-identification (anonymization) is not a robust technology for big data in the face of data fusion; in some instances, there may be compelling reasons to retain data for beneficial purposes. This right should be about use rather than collection. It should emphasize utilizing best practices to prevent inappropriate use of data during the data's whole life cycle, rather than depending on de-identification. It should not depend on a company's being able itself to recognize "all" the data about a consumer that it holds, which is increasingly technically infeasible.

The principles underlying CPBR's Security and Accountability remain valid in a use-based regime. They need to be applied throughout the value chain that includes data collection, analysis, and use.

Turn next to the rights here classified as consumer empowerments.

Where consumer empowerments have become practically impossible for the consumer to exercise meaningfully, they need to be recast as obligations of the commercial entity that actually uses the data or products of data analysis. This applies to the CPBR's principles of Individual Control and of Transparency.

Section 4.3 explained how the non-obvious nature of big data's products of analysis make it all but impossible for an individual to make fine-grained privacy choices for every new situation or app. For the principle of Individual Control to have meaning, PCAST believes that the burden should no longer fall on the consumer to manage privacy for each company with which the consumer interacts by a framework like "notice and consent." Rather, each company should take responsibility for conforming its uses of personal data to a personal privacy profile designated by the consumer and made available to that company (including from a third party designated by the consumer). Section 4.5.1 proposed a mechanism for this change in responsibility.

Transparency (in the sense of disclosure of privacy practices) suffers from many of the same problems. Today, the consumer receives an unhelpful blizzard of privacy-policy notifications, many of which say, in essence, "we providers can do anything we want."[130] As with Individual Control, the burden of conforming to a consumer's stated personal-privacy profile should fall on the company, with notification to the consumers by a company if their profile precludes that company's accepting their business. Since companies do not like to lose business, a positive market dynamic for competing privacy practices would thus be created.

For the right of Access and Accuracy to be meaningful, personal data must include the fruits of data analytics, not just collection. However, as this report has already explained (Section 4.4.2), it is not always possible for a company to "know what it knows" about a consumer, since that information may be unrecognized in the data; or it may become identifiable only in the future, when data sets are combined using new algorithms. When, however, the personal character of data is apparent to a company by virtue of its use of the data, its obligation to provide means for the correction of errors should be triggered. Consumers should have an expectation that companies will validate and correct data stemming from analysis and, since not all errors will be corrected, will also take steps to minimize the risk of adverse consequences to consumers from the use of inaccurate data. Again, the primary burden must fall on the commercial user of big data and not on the consumer.

5. PCAST PERSPECTIVES AND CONCLUSIONS

Breaches of privacy can cause harm to individuals and groups. It is a role of government to prevent such harm where possible, and to facilitate means of redress when the harm occurs. Technical enhancements of privacy can be effective only when accompanied by regulations or laws because, unless some penalties are enforced, there is no end to the escalation of the measures-countermeasures "game" between violators and protectors. Rules and regulations provide both deterrence of harmful actions and incentives to deploy privacy-protecting software technologies.

From everything already said, it should be obvious that new sources of big data are abundant; that they will continue to grow; and that they can bring enormous economic and social benefits. Similarly, and of comparable importance, new algorithms, software, and hardware technologies will continue to increase the power of data analytics in unexpected ways. Given these new capabilities of data aggregation and processing, there is inevitably new potential for both the unintentional leaking of both bulk and fine-grained data about individuals, and for new systematic attacks on privacy by those so minded.

Cameras, sensors, and other observational or mobile technologies raise new privacy concerns. Individuals often do not knowingly consent to providing data. These devices naturally pull in data unrelated to their primary purpose. Their data collection is often invisible. Analysis technology (such as facial, scene, speech, and voice recognition technology) is improving rapidly. Mobile devices provide location information that might not be otherwise volunteered. The combination of data from those sources can yield privacy-threatening information unbeknownst to the affected individuals.

It is also true, however, that privacy-sensitive data cannot always be reliably recognized when they are first collected, because the privacy-sensitive elements may be only latent in the data, made visible only by analytics (including those not yet invented), or by fusion with other data sources (including those not yet known). Suppressing the collection of privacy-sensitive data would thus be increasingly difficult, and it would also be increasingly counterproductive, frustrating the development of big data's socially important and economic benefits.

Nor would it be desirable to suppress the combining of multiple sources and kinds of data: Much of the power of big data stems from this kind of data fusion. That said, it remains a matter of concern that considerable amounts of personal data may be derived from data fusion. In other words, such data can be obtained or inferred without intentional personal disclosure.

It is an unavoidable fact that particular collections of big data and particular kinds of analysis will often have both beneficial and privacy-inappropriate uses. The appropriate use of both the data and the analyses are highly contextual.

Any specific harm or adverse consequence is the result of data, or their analytical product, passing through the control of three distinguishable classes of actor in the value chain:

First, there are *data collectors*, who control the interfaces to individuals or to the environment. Data collectors may collect data from clearly private realms (e.g., a health questionnaire or wearable sensor), from ambiguous situations (e.g., cell-phone pictures or Google Glass videos taken at a party or cameras and microphones placed in a classroom for remote broadcast), or – increasing in both quantity and quality – data from the "public square," where privacy-sensitive data may be latent and initially unrecognizable.

Second, there are *data analyzers*. This is where the "big" in big data becomes important. Analyzers may aggregate data from many sources, and they may share data with other analyzers. Analyzers, as distinct from collectors, create uses ("products of analysis") by bringing together algorithms and data sets in a large-scale computational environment. Importantly, analyzers are the locus where individuals may be profiled by data fusion or statistical inference.

Third, there are *users of the analyzed data* – business, government, or individual. Users will generally have a commercial relationship with analyzers; they will be purchasers or licensees (etc.) of the analyzer's products of analysis. It is the user who creates desirable economic and social outcomes. But, it is also the user who is the locus of producing actual adverse consequences or harms, when such occur.

5.1. Technical Feasibility of Policy Interventions

Policy, as created by new legislation or within existing regulatory authorities, can, in principle, intervene at various stages in the value chain described above. Not all such interventions are equally feasible from a technical perspective, or equally desirable if the societal and economic benefits of big data are to be realized.

As indicated in Section 4, basing policy on the control of collection is unlikely to succeed, except in very limited circumstances where there is an explicitly private context (e.g., measurement or disclosure of health data) and the possibility of *meaningful* explicit or implicit notice and consent (e.g., by privacy preference profiles, see Sections 4.3 and 4.5.1), which does not exist today.

There is little technical likelihood that "a right to forget" or similar limits on retention could be meaningfully defined or enforced (see Section 4.4.2). Increasingly, it will not be technically possible to surface "all" of the data about an individual. Policy based on protection by anonymization is futile, because the feasibility of re-identification increases rapidly with the amount of additional data (see Section 4.4.1). There is little, and decreasing, meaningful distinction between data and metadata. The capabilities of data fusion, data mining, and re-identification render metadata not much less problematic than data (see Section 3.1).

Even if direct controls on collection are in most cases infeasible, however, attention to collection practices may help to reduce risk in some circumstances. Such best practices as tracking provenance, auditing access and use, and continuous monitoring and control (see Sections 4.5.2 and 4.5.3) could be driven by partnerships between government and industry (the carrot) and also by clarifying tort law and defining what might constitute negligence (the stick).

Turn next to data analyzers. One the one hand, it may be difficult to regulate them, because their actions do not directly touch the individual (it is neither collection nor use) and may have no external visibility. Mere inference about an individual, absent its publication or use, may not be a feasible target of regulation. On the other hand, an increasing fraction of privacy issues will surface only with the application of data analytics. Many privacy challenges will arise from the analysis of data collected unintentionally that was not, at the time of collection, targeted at any particular individual or even group of individuals. This is because combining data from many sources will become more and more powerful.

It might be feasible to introduce regulation at the "moment of particularization" of data about an individual, or when this is done for some minimum number of individuals concurrently. To be effective such regulation would need to be accompanied by requirements for tracking provenance, auditing access and use, and using security measures (e.g., robust encryption infrastructure) at all stages of the evolution of data, and for providing transparency, and/or notification, at the moment of particularization.

Big data's "products of analysis" are created by computer programs that bring together algorithms and data so as to produce something of value. It might be feasible to recognize such programs, or their products, in a legal sense and to regulate their commerce. For example, they might not be allowed to be used in commerce (sold, leased, licensed, and so on) unless they are consistent with individuals' privacy elections or other expressions of community values (see Sections 4.3 and 4.5.1). Requirements might be imposed on conformity to appropriate standards of provenance, auditability, accuracy, and so on, in the data they use and produce; or that they meaningfully identify who (licensor vs. licensee) is responsible for correcting errors and liable for various types of harm or adverse consequence caused by the product.

It is not, however, the mere development of a product of analysis that can cause adverse consequences. Those occur only with its actual use, whether in commerce, by government, by the press, or by individuals. This seems the most technically feasible place to apply regulation going forward, focusing at the locus where harm can be produced, not far upstream from where it may barely (if at all) be identifiable.

When products of analysis produce imperfect information that may misclassify individuals in ways that produce adverse consequences, one might require that they meet standards for data accuracy and integrity; that there are useable interfaces that allow an individual to correct the record with voluntary additional information; and that there exist streamlined options for redress, including financial redress, when adverse consequences reach a certain level.

Some harms may affect groups (e.g., the poor or minorities) rather than identifiable individuals. Mechanisms for redress in such cases need to be developed.

There is a need to clarify standards for liability in case of adverse consequences from privacy violations. Currently there is a patchwork of out-of-date state laws and legal precedents. One could encourage the drafting of technologically savvy model legislation on cyber-torts for consideration by the states.

Finally, government may be forbidden from certain classes of uses, despite their being available in the private sector.

5.2. Recommendations

PCAST's charge for this study does not ask it to make recommendations on privacy policies, but rather to make a relative assessment of the technical feasibility of different broad policy approaches. PCAST's overall conclusions about that question are embodied in the first two of our recommendations:

Recommendation 1. Policy attention should focus more on the actual uses of big data and less on its collection and analysis.

By actual uses, we mean the specific events where something happens that can cause an adverse consequence or harm to an individual or class of individuals. In the context of big data, these events ("uses") are almost always actions of a computer program or app interacting either with the raw data or with the fruits of analysis of those data. In this formulation, it is not the data themselves that cause the harm, nor the program itself (absent any data), but the confluence of the two. These "use events" (in commerce, by government, or by individuals) embody the necessary specificity to be the subject of regulation. Since the purpose of bringing program and data together is to accomplish some identifiable desired task, use events also capture some notion of intent, in a way that data collection by itself or program development by itself may not. The policy question of what kinds of adverse consequences or harms rise to the level of needing regulation is outside of PCAST's charge, but an illustrative set that seem grounded in common American values was provided in Section 1.4.

PCAST judges that alternative big-data policies that focus on the regulation of data collection, storage, retention, a priori limitations on applications, and analysis (absent identifiable actual uses of big data or its products of analysis) are unlikely to yield effective strategies for improving privacy. Such policies are unlikely to be scalable over time as it becomes increasingly difficult to ascertain, about any particular data set, what personal information may be latent in it – or in its possible fusion with every other possible data set, present or future. The related issue is that policies limiting collection and retention are increasingly unlikely to be enforceable by other than severe and economically damaging measures. While there are certain definable classes of data so repugnant to society that their mere possession is criminalized,[131] the information in big data that may raise privacy concerns is increasingly inseparable from a vast volume of the data of ordinary commerce, or government function, or collection in the public square. This dual-use character of information, too, argues for the regulation of use rather than collection.

Recommendation 2. Policies and regulation, at all levels of government, should not embed particular technological solutions, but rather should be stated in terms of intended outcomes.

To avoid falling behind the technology, it is essential that policy concerning privacy protection should address the purpose (the "what") rather than the mechanism (the "how"). For example, regulating disclosure of health information by regulating the use of anonymization fails to capture the power of data fusion; regulating the protection of information about minors by controlling inspection of student records held by schools fails to anticipate the student information capturing by online learning technologies. Regulating control of the inappropriate disclosure of health information or student performance, no matter how the data are acquired is more robust.

PCAST further responds to its charge with the following recommendations, intended to advance the agenda of strong privacy values and the technological tools needed to support them:

Recommendation 3. With coordination and encouragement from OSTP, the NITRD agencies[132] should strengthen U.S. research in privacy-

related technologies and in the relevant areas of social science that inform the successful application of those technologies.

Some of the technology for controlling uses already exists. Research (and funding for it) is needed, however, in the technologies that help to protect privacy, in the social mechanisms that influence privacy-preserving behavior, and in the legal options that are robust to changes in technology and create appropriate balance among economic opportunity, other national priorities, and privacy protection.

Following up on recommendations from PCAST for increased privacy-related research,[133] a 2013-2014 internal government review of privacy-focused research across Federal agencies supporting research on information technologies suggests that about $80 million supports either research with an explicit focus on enhancing privacy or research that addresses privacy protection ancillary to some other goal (typically cybersecurity).[134] The funded research addresses such topics as an individual's control over his or her information, transparency, access and accuracy, and accountability. It is typically of a general nature, except for research focusing on the health domain or (relatively new) consumer energy usage. The broadest and most varied support for privacy research, in the form of grants to individuals and centers, comes from the National Science Foundation (NSF), engaging social science as well as computer science and engineering.[135,136]

Research into privacy as an extension or complement to security is supported by a variety of Department of Defense agencies (Air Force Research Laboratory, the Army's Telemedicine and Advanced Technology Research Center, Defense Advanced Research Projects Agency, National Security Agency, and Office of Naval Research) and the Intelligence Advanced Research Projects Activity (IARPA) within the Intelligence Community. IARPA, for example, has hosted the Security and Privacy Assurance Research[137] program, which has explored a variety of encryption techniques. Research at the National Institute for Standards and Technology (NIST) focuses on the development of cryptography and biometric technology to enhance privacy as well as support for federal standards and programs for identity management.[138]

Looking to the future, continued investment is needed not only in privacy topics ancillary to security, but also in automating privacy protection for the broadest aspects of use of data from all sources. Relevant topics include cryptography, privacy-preserving data mining (including analysis of streaming as well as stored) data,[139] formalization of privacy policies, tools for automating conformance of software to personal privacy policy and to legal policy, methods for auditing use in context and identifying violations of policy, and research on enhancing people's ability to make sense of the results of various big-data analyses. Development of technologies that support both quality analytics and privacy preservation on distributed data, such as secure multiparty computation, will become even more important, given the expectation that people will draw increasingly from data stored in multiple locations. The creation of tools that analyze the panoply of National, state, regional, and international rules and regulations for inconsistencies and differences will be helpful for the definition of new rules and regulations, as well as for those software developers that need to customize their services for different markets.

Recommendation 4. OSTP, together with the appropriate educational institutions and professional societies, should encourage increased education and training opportunities concerning privacy protection, including professional career paths.

Programs that provide education leading to privacy expertise (akin to what is being done for security expertise) are essential and need encouragement. One might envision careers for digital-privacy experts both on the software development side and on the technical management side. Employment opportunities should exist not only in industry (and government at all levels), where jobs focused on privacy (including but not limited to Chief Privacy Officers) have been growing, but also for consumer and citizen advocacy and support, perhaps offering "annual privacy checkups" for individuals. Just as education and training about cybersecurity has advanced over the past 20 years within the technical community, there is now opportunity to educate and train students about privacy implications and privacy enhancements, beyond the present small niche area occupied by this focus within computer science programs.[140] Privacy is also an important component of ethics education for technology professionals.

Recommendation 5. The United States should take the lead both in the international arena and at home by adopting policies that stimulate the use of practical privacy-protecting technologies that exist today. This country can exhibit leadership both by its convening power (for instance, by promoting the creation and adoption of standards) and also by its own procurement practices (such as its own use of privacy-preserving cloud services).

Section 4.5.2 described a set of privacy-enhancing best practices that already exist today in U.S. markets. PCAST is not aware of any more effective innovation or strategies being developed abroad; rather, some countries seem inclined to pursue what PCAST believes to be blind alleys. This circumstance offers an opportunity for U.S. technical leadership in privacy in the international arena, an opportunity that should be seized. Public policy can help to nurture the budding commercial potential of privacy-enhancing technologies, both through U.S. government procurement and through the larger policy framework that motivates private-sector technology engagement.

As it does for security, cloud computing offers positive new opportunities for privacy. By requiring privacy- enhancing services from cloud-service providers contracting with the U. S. government, the government should encourage those providers to make available sophisticated privacy enhancing technologies to small businesses and their customers, beyond what the small business might be able to do on its own.[141]

5.3. Final Remarks

Privacy is an important human value. The advance of technology both threatens personal privacy and provides opportunities to enhance its protection. The challenge for the U.S. Government and the larger community, both within this country and globally, is to understand what the nature of privacy is in the modern world and to find those technological, educational, and policy avenues that will preserve and protect it.

APPENDIX A. ADDITIONAL EXPERTS PROVIDING INPUT

Yochai Benkler
Harvard
Eleanor Birrell
Cornell University
Courtney Bowman
Palantir
Christopher Clifton
Purdue University
James Costa
Sandia National Laboratory
Lorrie Faith Cranor

Peter Guerra
Booz Allen
Michael Jordan
University of California, Berkeley
Philip Kegelmeyer
Sandia National Laboratory
Angelos Keromytis
Columbia University
Thomas Kalil
OSTP
Jon Kleinberg

Carnegie Mellon University
Deborah Estrin
Cornell NYC
William W. (Terry) Fisher
Harvard Law School
Stephanie Forrest
University of New Mexico
Dan Geer
In-Q-Tel
Deborah K. Gracio
Pacific Northwest National Laboratory
Eric Grosse
Google
Deirdre Mulligan
University of California, Berkeley
Leonard Napolitano
Sandia National Laboratory
Charles Nelson
OSTP

Chris Oehmen
Pacific Northwest National Laboratory
Alex "Sandy" Pentland
Massachusetts Institute of Technology
Rene Peralta
National Institute of Standards and Technology
Anthony Philippakis
Genome Bridge

Timothy Polk
OSTP
Fred B. Schneider
Cornell University
Greg Shipley
In-Q-Tel

Cornell University
Julia Lane
American Institutes for Research
Carl Landwehr
George Washington University
David Moon
Ernst & Young
Keith Marzullo
National Science Foundation
Martha Minow
Harvard Law School

Tom Mitchell
Carnegie Mellon University
Lauren Smith
OSTP
Francis Sullivan
Institute for Defense Analysis
Thomas Vagoun
NITRD National Coordination Office
Konrad Vesey
Intelligence Advanced Research Activity
James Waldo
Harvard

Peter Weinberger
Google, Inc.

Daniel J. Weitzner
Massachusetts Institute of Technology
Nicole Wong
OSTP
Jonathan Zittrain
Harvard Law School

Special Acknowledgment

PCAST is especially grateful for the rapid and comprehensive assistance provided by an ad hoc group of staff at the National Science Foundation (NSF), Computer and Information Science and Engineering Directorate. This team was led by Fen Zhao and Emily Grumbling, who were enlisted by Suzanne Iacono. Drs. Zhao and Grumbling worked tirelessly to review the technical literature, elicit perspectives and feedback from a range of NSF colleagues, and iterate on descriptions of numerous technologies relevant to big data and privacy and how those technologies were evolving.

NSF Technology Team Leaders
    Fen Zhao, AAAS Fellow, CISE
    Emily Grumbling, AAAS Fellow, Office of Cyberinfrastructure

Additional NSF Contributors
    Robert Chadduck, Program Director
    Almadena Y. Chtchelkanova, Program Director
    David Corman, Program Director
    James Donlon, Program Director
    Jeremy Epstein, Program Director
    Joseph B. Lyles, Program Director
    Dmitry Maslov, Program Director
    Mimi McClure, Associate Program Director
    Anita Nikolich, Expert
    Amy Walton, Program Director
    Ralph Wachter, Program Director

About the President's Council of Advisors on Science and Technology

The President's Council of Advisors on Science and Technology (PCAST) is an advisory group of the Nation's leading scientists and engineers, appointed by the President to augment the science and technology advice available to him from inside the White House and from cabinet departments and other Federal agencies. PCAST is consulted about, and often makes policy recommendations concerning, the full range of issues where understandings from the domains of science, technology, and innovation bear potentially on the policy choices before the President.

End Notes

[1] The White House Office of Science and Technology Policy
[2] NITRD refers to the Networking and Information Technology Research and Development program, whose participating Federal agencies support unclassified research in advanced information technologies such as computing, networking, and software and include both research- and mission-focused agencies such as NSF, NIH, NIST, DARPA, NOAA, DOE's Office of Science, and the D0D military-service laboratories (see http://www.nitrd.gov/SUBCOMMITTEE/nitrd_agencies/index.aspx).
[3] "Remarks by the President on Review of Signals Intelligence," January 17, 2014. http://www.whitehouse.gov/the-press-office/2014/01/17/remarks-president-review-signals-intelligence
[4] Gartner, Inc., "IT Glossary." https://www.gartner.com/it-glossary/big-data/
[5] Barker, Adam and Jonathan Stuart Ward, "Undefined By Data: A Survey of Big Data Definitions," arXiv:1309.5821. http://arxiv.org/abs/1309.5821
[6] PCAST acknowledges gratefully the assistance of several contributors at the National Science Foundation, who helped to identify and distill key insights from the technical literature and research community, as well as other technical experts in academia and industry that it consulted during this project. See Appendix A.
[7] Seipp, David J., *The Right to Privacy in American History*, Harvard University, Program on Information Resources Policy, Cambridge, MA, 1978.
[8] Warren, Samuel D. and Louis D. Brandeis, "The Right to Privacy." *Harvard Law Review* 4:5, 193, December 15, 1890.
[9] Id. at 195.
[10] Digital Media Law Project, "Publishing *Personal* and Private Information." http://www.dmlp.org/legal-guide/publishing-personal-and-private-information
[11] *Griswold v. Connecticut*, 381 U.S. 479 (1965).
[12] Id. at 483-84.
[13] *Olmstead v. United States*, 277 U.S. 438 (1928).
[14] *McIntyre v. Ohio Elections Commission*, 514 U.S. 334, 340-41 (1995). The decision reads in part, *"Protections for anonymous speech are vital to democratic discourse. Allowing dissenters to shield their identities frees them to express critical minority views . . . Anonymity is a shield from the tyranny of the majority. . . . It thus exemplifies the purpose behind the Bill of Rights and of the First Amendment in particular: to protect unpopular individuals from retaliation . . . at the hand of an intolerant society."*
[15] Federal Trade Commission, "Privacy Online: Fair Information Practices in the Electronic Marketplace," May 2000.
[16] Genetic Information Nondiscrimination Act of 2008, PL 110–233, May 21, 2008, 122 Stat 881.
[17] One Hundred Tenth Congress, "Privacy: The use of commercial information resellers by federal agencies," *Hearing before the Subcommittee on Information Policy, Census, and National Archives of the Committee on Oversight and Government Reform*, House of Representatives, March 11, 2008.
[18] For example, Experian provides much of Healthcare.gov's identity verification component using consumer credit information not available to the government. See *Consumer Reports*, "Having trouble proving your identity to HealthCare.gov? Here's how the process works," December 18, 2013. http://www.consumerreports.org/cro/news/2013/12/how-to-prove-your-identity-on-healthcare- gov/index.htm?loginMethod=auto
[19] Warren, Samuel D. and Louis D. Brandeis, "The Right to Privacy." *Harvard Law Review* 4:5, 193, December 15, 1890.
[20] Prosser, William L., "Privacy," *California Law Review* 48:383, 389, 1960.
[21] Id.

[22] (1) Digital Media Law Project, "Publishing Personal and Private Information." http://www.dmlp.org/legal- guide/publishing-personal-and-private-information. (2) Id., "Elements of an Intrusion Claim." http://www.dmlp.org/legal- guide/elements-intrusion-claim

[23] One perspective informed by new technologies and technology-medicated communication suggests that privacy is about the "continual management of boundaries between different spheres of action and degrees of disclosure within those spheres," with privacy and one's public face being balanced in different ways at different times. See: Leysia Palen and Paul Dourish, "Unpacking 'Privacy' for a Networked World," *Proceedings of CHI 2003*, Association for Computing Machinery, April 5-10, 2003.

[24] "I would ask whether people reasonably expect that their movements will be recorded and aggregated in a manner that enables the Government to ascertain, more or less at will, their political and religious beliefs, sexual habits, and so on." *United States v. Jones* (10-1259), Sotomayor concurrence at http://www.supremecourt.gov/opinions/11pdf/10-1259.pdf.

[25] Dick, Phillip K., "The Minority Report," first published in *Fantastic Universe* (1956) and reprinted in *Selected Stories of Philip K. Dick*, New York: Pantheon, 2002.

[26] ElBoghdady, Dina, "Advertisers Tune In to New Radio Gauge," *The Washington Post*, October 25, 2004. http://www.washingtonpost.com/wp-dyn/articles/A60013-2004Oct24.html

[27] American Civil Liberties Union, "You Are Being Tracked: How License Plate Readers Are Being Used To Record Americans' Movements," July, 2013. https://www.aclu.org/files/assets-aclu-alprreport-opt-v05.pdf

[28] Hardy, Quentin, "How Urban Anonymity Disappears When All Data Is Tracked," *The New York Times*, April 19, 2014.

[29] Rudin, Cynthia, "Predictive policing: Using Machine Learning to Detect Patterns of Crime," *Wired*, August 22, 2013. http://www.wired.com/insights/2013/08/predictive-policing-using-machine-learning-to-detect-patterns-of-crime/.

[30] (1) Schiller, Benjamin, "First Degree Price Discrimination Using Big Data," Jan. 30. 2014, Brandeis University. http://benjaminshiller.com/imagesand http://www.forbes.com/sites/modeledbehavior/2013/09/01/ will-big-data-bring-more-price-discrimination/ (2) Fisher, William W. "When Should We Permit Differential Pricing of Information?" *UCLA Law Review* 55:1, 2007.

[31] Burn-Murdoch, John, "UK technology firm uses machine learning to combat gambling addiction," *The Guardian*, August 1, 2013. http://www.theguardian.com/news/datablog/2013/aug/01/uk-firm-uses-machine-learning-fight-gambling-addiction

[32] Clifford, Stephanie, "Using Data to Stage-Manage Paths to the Prescription Counter," *The New York Times*, June 19, 2013. http://bits.blogs.nytimes.com/2013/06/19/using-data-to-stage-manage-paths-to-the-prescription-counter/

[33] Clifford, Stephanie, "Attention, Shoppers: Store Is Tracking Your Cell," *The New York Times*, July 14, 2013.

[34] Duhigg, Charles, "How Companies Learn Your Secrets," *The New York Times Magazine*, February 12, 2012. http://www.nytimes.com/2012/02/19/magazine/shopping-habits.html?pagewanted=all&_r=0

[35] Volokh, Eugene, "Outing Anonymous Bloggers," June 8, 2009. http://www.volokh.com/2009/06/08/outing-anonymous-bloggers/; A. Narayanan et al., "On the Feasibility of Internet-Scale Author Identification," IEEE Symposium on Security and Privacy, May 2012. http://ieeexplore.ieee.org/stamp/stamp.jsp?tp=&arnumber=6234420

[36] Facebook's "The Graph API" (at https://developers.facebook.com/docs/graph-api/) describes how to write computer programs that can access the Facebook friends' data.

[37] One of four big-data applications honored by the trade journal, *Computerworld*, in 2013. King, Julia, "UN tackles socio-economic crises with big data," *Computerworld*, June 3, 2013. http://www.computerworld.com/s/article/print/9239643/UN_tackles_socio_economic_crises_with_big_data

Big Data and Privacy: A Technological Perspective 163

[38] Ungerleider, Neal, "This May Be The Most Vital Use Of "Big Data" We've Ever Seen," *Fast Company*, July 12, 2013. http://www.fastcolabs.com/3014191/this-may-be-the-most-vital-use-of-big-data-weve-ever-seen.

[39] Center for Data Innovations, *100 Data Innovations*, Information Technology and Innovation Foundation, Washington, DC, January 2014. http://www2.datainnovation.org/2014-100-data-innovations.pdf

[40] Waters, Richard, "Data open doors to financial innovation," *Financial Times*, December 13, 2013. http://www.ft.com/intl/cms/s/2/3c59d58a-43fb-11e2-844c-00144feabdc0.html

[41] (1) Wiens, Jenna, John Guttag, and Eric Horvitz, "A Study in Transfer Learning: Leveraging Data from Multiple Hospitals to Enhance Hospital-Specific Predictions," *Journal of the American Medical Informatics* Association, January 2014. (2) Weitzner, Daniel J., et al., "Consumer Privacy Bill of Rights and Big Data: Response to White House Office of Science and Technology Policy Request for Information," April 4, 2014.

[42] Frazer, Bryant, "MIT Computer Program Reveals Invisible Motion in Video," *The New York Times* video, February 27, 2013. https://www.youtube.com/watch?v=3rWycBEHn3s

[43] For an overview of MOOCs and associated analytics opportunities, see PCAST's December 2013 letter to the President. http://www.whitehouse.gov/sites/default/files/microsites/ostp/PCAST/pcast_edit_dec-2013.pdf

[44] There is also uncertainty about how to interpret applicable laws, such as the Family Educational Rights and Privacy Act (FERPA). Recent Federal guidance is intended to help clarify the situation. See: U.S. Department of Education, "**Protecting Student Privacy While Using Online Educational Services: Requirements and Best Practices**," February 2014. http://ptac.ed.gov/sites/default/files/Student%20Privacy%20and%20Online%20Educational%20Services%20%28February%202014%29.pdf

[45] Cukier, Kenneth, and Viktor Mayer-Schoenberger, "How Big Data Will Haunt You Forever," *Quartz*, March 11, 2014. http://qz.com/185252/how-big-data-will-haunt-you-forever-your-high-school-transcript/

[46] Nest, acquired by Google, attracted attention early for its design and its use of big data to adapt to consumer behavior. See: Aoki, Kenji, "Nest Gives the Lowly Smoke Detector a Brain," *Wired*, October, 2013. http://www.wired.com/2013/10/nest-smoke-detector/all/

[47] Reuters, "Apple acquires Israeli 3D chip developer PrimeSense," November 25, 2013. http://www.reuters.com/article/2013/11/25/us-primesense-offer-apple-idUSBRE9AO04C20131125

[48] Id.

[49] Google, "Glass gestures." https://support.google.com/glass/answer/3064184?hl=en

[50] Tene, Omer, and Jules Polonetsky, "A Theory of Creepy: Technology, Privacy and Shifting Social Norms," *Yale Journal of Law and Technology* 16:59, 2013, pp. 59-100.

[51] See references at footnote 30.

[52] Such databases endure and form the basis of continuing concern among privacy advocates.

[53] Schemas are formal definitions of the configuration of a database: its tables, relations, and indices. Headers are the sometimes-invisible prefaces to email messages that contain information about the sending and destination addresses and sometimes the routing of the path between them.

[54] In the Internet and similar networks, information is broken up into chunks called packets, which may travel independently and depend on metadata to be reassembled properly at the destination of the transmission.

[55] Federal Trade Commission, "FTC Staff Revises Online Behavioral Advertising Principles," Press Release, February 12, 2009. http://www.ftc.gov/news-events/press-releases/2009/02/ftc-staff-revises-online-behavioral-advertising-principles

[56] (1) Cf. *The Wall Street Journal*'s "What they know" series (http://online.wsj.com/public/page/what-they-know-digital-privacy.html). (2) Turow, Joseph, *The Daily You: How the Advertising Industry is Defining your Identity and Your Worth*, Yale University Press, 2012. http://yalepress.yale.edu/book.asp?isbn=9780300165012

[57] DuckDuckGo is a non-tracking search engine that, while perhaps yielding fewer results than leading search engines, is used by those looking for less tracking. See: https://duckduckgo.com/

[58] (1) Tanner, Adam, "The Web Cookie Is Dying. Here's The Creepier Technology That Comes Next," *Forbes*, June 17, 2013. http://www.forbes.com/sites/adamtanner/2013/06/17/the-web-cookie-is-dying-heres-the-creepier-technology-that- comes-next/ (2) Acar, G. et al., "FPDetective: Dusting the Web for Fingerprinters," 2013. http://www.cosic.esat.kuleuven.be/publications/article-2334.pdf

[59] Federal Trade Commission, "Android Flashlight App Developer Settles FTC Charges It Deceived Consumers," *Press Release*, December 5, 2013. http://www.ftc.gov/news-events/press-releases/2013/12/android-flashlight-app-developer- settles-ftc-charges-it-deceived

[60] (1) FTC File No. 132-3087 Decision and order. http://www.ftc.gov/system/files/documents/cases/140409goldenshoresdo.pdf >(2) "FTC Approves Final Order Settling Charges Against Flashlight App Creator." http://www.ftc.gov/news-events/press-releases/2014/04/ftc-approves-final- order-settling-charges-against-flashlight-app

[61] See: http://www.fitbit.com/

[62] Koonin, Steven E., Gregory Dobler and Jonathan S. Wurtele, "Urban Physics," *American Physical Society News*, March, 2014. http://www.aps.org/publications/apsnews/201403/urban.cfm

[63] Durand, Fredo, et al., "MIT Computer Program Reveals Invisible Motion in Video," *The New York Times*, video, February 27, 2013. https://www.youtube.com/watch?v=3rWycBEHn3s

[64] Feldman, Ronen, "Techniques and Applications for Sentiment Analysis," *Communications of the ACM*, 56:4, pp. 82-89.

[65] Mayer-Schönberger, Viktor and Kenneth Cukier, *Big Data: A Revolution That Will Transform How We Live, Work, and Think,* Boston, NY: Houghton Mifflin Harcourt, 2013.

[66] National Research Council, *Frontiers in Massive Data* Analysis, National Academies Press, 2013.

[67] (1) Thill, Brent and Nicole Hayashi, *Big Data = Big Disruption: One of the Most Transformative IT Trends Over the Next Decade*, UBS Securities LLC, October 2013. (2) McKinsey Global Institute, Center for Government, and Business Technology Office, *Open data: Unlocking innovation and performance with liquid information*, McKinsey & Company, October 2013.

[68] Le, Q.V. et al., "Building High-level Features Using Large Scale Unsupervised Learning," http://static.googleusercontent.com/media/research.google.com/en/us/archive/unsupervised_icml2012.pdf

[69] Bramer, M., "Principles of Data Mining," *Springer*, 2013.

[70] Mitchell, Tom M., "The Discipline of Machine Learning," Technical Report CMU-ML-06-108, Carnegie Mellon University, July 2006.

[71] DARPA, for example, has a project involving machine learning and other technologies to build medical causal models from analysis of cancer literature, leveraging the greater capacity of a computer than a person to process information from a large number of sources. See description at http://www.darpa.mil/Our_Work/I2O/Programs/Big_Mechanism.aspx

[72] "Data mining breaks the basic intuition that identity is the greatest source of potential harm because it substitutes inference for identifying information as a bridge to get at additional facts." Barocas, Solon and Helen Nissenbaum, "Big Data's End Run Around Anonymity and Consent," Chapter II, in Lane, Julia, et al., *Privacy, Big Data, and the Public Good,* Cambridge University Press, 2014.

[73] Manyika, J. et al., "Big Data: The next frontier for innovation, competition, and productivity," *McKinsey Global Institute*, 2011.

[74] Navarro-Arriba, G. and V. Torra, "Information fusion in data privacy: A survey," *Information Fusion*, 13:4, 2012, pp. 235-244.

[75] Khaleghi, B. et al., "Multisensor data fusion: A review of the state-of-the-art," *Information Fusion*, 14:1, 2013, pp. 28-44.

[76] Lam, J., et al., "Urban scene extraction from mobile ground based lidar data," *Proceedings of 3DPVT*, 2010.

[77] Agarwal, S., et al., "Building Rome in a day," *Communications of the ACM*, 54:10, 2011, pp. 105-112.

[78] Workshop on Frontiers in Image and Video Analysis, National Science Foundation, Federal Bureau of Investigation, Defense Advanced Research Projects Agency, and University of Maryland Institute for Advanced Computer Studies, January 28-29, 2014. http://www.umiacs.umd.edu/conferences/fiva/

[79] For example, Newark Airport recently installed a system of 171 LED lights (from Sensity [http://www.sensity.com/]) that contain special chips to connect to sensors and cameras over a wireless system. These systems allow for advanced automatic lighting to improve security in places like parking garages, and in doing so capture a large range of information.

[80] This was discussed at the workshop cited in footnote 78.

[81] Such concerns are likely to grow as commercial satellite imagery systems such as Skybox (http://skybox.com/) provide the basis for more services.

[82] Billitteri, Thomas J., et al. "Social Media Explosion: Do social networking sites threaten privacy rights?" *CQ Researcher*, January 25, 2013, 23:84-104.

[83] Juang, B.H. and Lawrence R. Rabiner, "Automated Speech Recognition – A Brief History of the Technology Development," October 8, 2004. http://www.ece.ucsb.edu/Faculty/Rabiner/ece259/Reprints/354_LALI-ASRHistory-final-10-8.pdf

[84] "Where Speech Recognition is Going," *Technology Review*, May 29, 2012. http://www.kurzweilai.net/where-speech-recognition-is-going

[85] Wasserman, S. "Social network analysis: Methods and applications," *Cambridge University Press*, 8, 1994.

[86] See, for example: (1) Backstrom, Lars, et al., "Inferring Social Ties from Geographic Coincidences," *Proceedings of the National Academy of Sciences*, 2010. (2) Backsrom, Lars, et al., "Wherefore Art Though R3579X? Anonymized Social Networks, Hidden Patterns, and Structural Steganography," *International World Wide Web Conference* 2007, Alberta, Canada, May 12, 2007.

[87] A variety of tools exist for managing, analyzing, visualizing and manipulating network (graph) datasets, such as Allegrograph, GraphVis, R, visone and Wolfram Alpha. Some, such as Cytoscape, Gephi and Netviz are open source.

[88] (1) Geetoor, L. and E. Zheleva, "Preserving the privacy of sensitive relationships in graph data," *Privacy, security, and trust in KDD*, 153-171, 2008. (2) Mislove, A., et al., "An analysis of social-based network Sybil defenses," *ACM SIGCOMM Computer Communication Review*, 2011. (3) Backstrom, Lars, et al., "Find Me If You Can: Improving Geographic Prediction with Social and Spatial Proximity," *Proceedings of the 19th international conference on World Wide Web*, 2010. (4) Backstrom, L. and J. Kleinberg, "Romantic Partnerships and the Dispersion of Social Ties: A Network Analysis of Relationship Status on Facebook," *Proceedings of the 17th ACM Conference on Computer Supported Cooperative Work and Social Computing* (CSCW), 2014.

[89] (1) Narayanan, A. and V. Shmatikov, "De-anonymizing social networks," *30th IEEE Symposium on Security and Privacy*, 173-187, 2009. (2) Crandall, David J., et al., "Inferring social ties from geographic coincidences," *Proceedings of the National Academy of Sciences*, 107:52, 2010. (3) Backstrom, L, C. Dwork and J. Kleinberg, "Wherefore Art Thou R3579X? Anonymized Social Networks, Hidden Patterns, and Structural Steganography," *Proceedings of the 16th Intl. World Wide Web Conference*, 2007. (4) Saramäki, Jari, et al., "Persistence of social signatures in human communication," *Proceedings of the National Academy of Sciences*, 111.3:942-947, 2014.

[90] Fienberg, S.E., "Is the Privacy of Network Data an Oxymoron?" *Journal of Privacy and Confidentiality*, 4:2, 2013.

[91] Krebs, V.E., "Mapping networks of terrorist cells," *Connections*, 24.3:43-52, 2002.

[92] Sundsøy, P. R., et al., "Product adoption networks and their growth in a large mobile phone network," *Advances in Social Networks Analysis and Mining (ASONAM)*, 2010.

[93] Hodgson, Bob, "A Vital New Marketing Metric: The Network Value of a Customer," *Predictive Marketing: Optimize Your ROI With Analytics*. http://predictive-marketing.com/index.php/a-vital-new-marketing-metric-the-network-value-of-a- customer/

[94] Backstrom, Lars et al, "Find me if you can: improving geographical prediction with social and spatial proximity," *Proceedings of the 19th international conference on World Wide Web*, 2010.

[95] "Top 20 social media monitoring vendors for business," *Socialmedia.biz*, http://socialmedia.biz/2011/01/12/top-20-social-media-monitoring-vendors-for-business/

[96] A petabyte is 10^{15} bytes. One petabyte could store the individual genomes of the entire U.S. population. The human brain has been estimated to have a capacity of 2.5 petabytes.

[97] McLellan, Charles, "The 21st Century Data Center: An Overview," *ZDNet*, April 2, 2013. http://www.zdnet.com/the-21st-century-data-center-an-overview-7000012996/

[98] See: http://accumulo.apache.org/

[99] See: https://amplab.cs.berkeley.edu/software/

[100] Cloud Security Alliance, "Big Data Working Group: Comment on Big Data and the Future of Privacy," March 2014. https://downloads.cloudsecurityalliance.org/initiatives/bdwg/Comment_on_Big_Data_Future_of_Privacy.pdf

[101] Qi, H. and A. Gani, "Research on mobile cloud computing: Review, trend and perspectives," *Digital Information and Communication Technology and it's Applications (DICTAP), 2012 Second International Conference on*, 2012.

[102] Jeffery, K. et al., "A vision for better cloud applications," *Proceedings of the 2013 International Workshop on Multi-Cloud Applications and Federated Clouds*, Prague, Czech Republic, MODAClouds, ACM Digital Library, April 22-23, 2013.

[103] PCAST has addressed issues in cybersecurity, both in reviewing the NITRD programs and directly in a 2013 report, *Immediate Opportunities for Strengthening the Nation's Cybersecurity*. http://www.whitehouse.gov/sites/default/files/microsites/ostp/PCAST/pcast_cybersecurity_nov-2013.pdf

[104] There are also choices in the design and implementation of security mechanisms that affect privacy. In particular, authentication or the attempt to demonstrate identity at some level can be done with varying degrees of disclosure. See, for example: Computer Science and Telecommunications Board, *Who Goes There: Authentication Through the Lens of Privacy*, National Academies Press, 2003.

[105] Such research can inform efforts to automate the checking of compliance with policies and/or associated auditing.

[106] This future-proofing remains hard to achieve; PCAST's cybersecurity report advocated approaches that would be more durable than the kinds of check-lists that are easily rendered obsolete. See: http://www.whitehouse.gov/sites/default/files/microsites/ostp/PCAST/pcast_cybersecurity_nov-2013.pdf

[107] See, for example: (1) Breaux, Travis D., and Ashwini Rao, "Formal Analysis of Privacy Requirements Specifications for Multi-Tier Applications," *21st IEEE Requirements Engineering Conference* (RE 2013), Rio de Janeiro, Brazil, July 2013. http://www.cs.cmu.edu/~agrao/paper/Analysis_of_Privacy_Requirements_Facebook(2) Feigenbaum, Joan, et al., "Towards a Formal Model of Accountability," *New Security Paradigms Workshop 2011*, Marin County, CA, September 12-15, 2011. http://www.nspw.org/papers/2011/nspw2011-feigenbaum.pdf

[108] Landwehr, Carl, "Engineered Controls for Dealing with Big Data," Chapter 10, in Lane, Julia, et al., *Privacy, Big Data, and the Public Good*, Cambridge University Press, 2014.

[109] The use of this term in computing originated with what is now viewed as a classic article: Brooks, Fred P., "No silver bullet – Essence and Accidents of Software Engineering", *IEEE Computer* 20:4, April 1987, pp. 10-19.

[110] Attacks that compromise the hardware or software that does the encrypting (for example, the promulgation of intentionally weak cryptography standards) can be considered to be a variant of attacks that reveal plaintext.

[111] "Krebs on Security, collected posts on Target data breach," 2014. http://krebsonsecurity.com/tag/target-data-breach/

[112] Public-key encryption originated through the secret work of British mathematicians at the U.K.'s Government Communications Headquarters (GCHQ), an organization roughly analogous to the NSA, and received broader attention through the independent work by researchers including Whitfield Diffie and Martin Hellman in the United States.

[113] Fisher, Dennis, "Final Report on DigiNotar Hack Shows Total Compromise of CA Servers," *ThreatPost*, October 31, 2012. http://threatpost.com/final-report-diginotar-hack-shows-total-compromise-ca-servers-103112/77170.

[114] It is not publicly known whether or not the earlier 2010 compromise of servers belonging to VeriSign, a much larger CA, led to compromises of certificates or signing authorities. Bradley, Tony, "VeriSign Hacked: What We Don't Know Might Hurt Us," *PC World*, February 2, 2012. http://www.pcworld.com/article/249242/verisign_hacked_what_we_dont_know_might_hurt_us.html

[115] A sample report-card: https://www.eff.org/deeplinks/2013/11/encrypt-web-report-whos-doing-what#crypto-chart

[116] Diffie, Whitfield, et al., "Authentication and Authenticated Key Exchanges" *Designs, Codes and Cryptography* 2:2, June 1992, pp.107-125.

[117] (1) Dwork, Cynthia, "Differential Privacy," 33rd International Colloquium on Automata, Languages and Programming, 2006. (2) Dwork, Cynthia, "A Firm Foundation for Private Data Analysis," *Communications of the ACM*, 54.1, 2011.

[118] Gindin, Susan E., "Nobody Reads Your Privacy Policy or Online Contract: Lessons Learned and Questions Raised by the FTC's Action against Sears," *Northwestern Journal of Technology and Intellectual Property* 1:8, 2009-2010.

[119] De-identification can also be seen as a spectrum, rather than a single approach. See: "Response to Request for Information Filed by U.S. Public Policy Council of the Association for Computing Machinery," March 2014.

[120] Sweeney, et al., "Identifying Participants in the Personal Genome Project by Name," *Harvard University Data Privacy Lab*. White Paper 1021-1, April 24, 2013. http://dataprivacylab.org/projects/pgp/

[121] See, for example: Ryan Whitwam, "Snap Save for iPhone Defeats the Purpose of Snapchat, Saves Everything Forever," *PC Magazine*, August 12, 2013. http://appscout.pcmag.com/apple-ios-iphone-ipad-ipod/314653-snap-save-for-iphone-defeats-the-purpose-of-snapchat-saves-everything-forever

[122] Abelson, Hal and Lalana Kagal, "Access Control is an Inadequate Framework for Privacy Protection," *W3C Workshop on Privacy for Advanced Web APIs 12/13*, July 2010, London. http://www.w3.org/2010/api-privacy-ws/papers.html

[123] Mundie, Craig, "Privacy Pragmatism: Focus on Data Use, Not Data Collection," *Foreign Affairs*, March/April, 2014.

[124] Nissenbaum, H., "Privacy in Context: Technology, Policy, and the Integrity of Social Life," *Stanford Law Books*, 2009.

[125] See references at footnote 107 and also: (1) Weitzner, D.J., et al., "Information Accountability," *Communications of the ACM*, June 2008, pp. 82-87. (2) Tschantz, Michael Carl, Anupam Datta, and Jeannette M. Wing, "Formalizing and Enforcing Purpose Restrictions in Privacy Policies." http://www.andrew.cmu.edu/user/danupam/TschantzDattaWing12.pdf

[126] For example, at Carnegie Mellon University, Lorrie Cranor directs the CyLab Usable Privacy and Security Laboratory (http://cups.cs.cmu.edu/). Also, see *2nd International Workshop on Accountability: Science, Technology and Policy*, MIT Computer Science and Artificial

Intelligence Laboratory, January 29-30, 2014. http://dig.csail.mit.edu/2014/AccountableSystems2014/

[127] Oracle's eXtensible Access Control Markup Language (XACML) has been used to implement attribute-based access controls for identity management systems. (Personal communication, Mark Gorenberg and Peter Guerra of Booz Allen)

[128] Office of the Director of National Intelligence, "IC CIO Enterprise Integration & Architecture: Trusted Data Format." http://www.dni.gov/index.php/about/organization/chief-information-officer/trusted-data-format

[129] See: http://www.openstack.org/

[130] Lawyers may encourage companies to use over-inclusive language to cover the unpredictable evolution of possibilities described elsewhere in this report, even in the absence of specific plans to use specific capabilities.

[131] Child pornography is the most universally recognized example.

[132] NITRD refers to the Networking and Information Technology Research and Development program, whose participating Federal agencies support unclassified research in in advanced information technologies such as computing, networking, and software and include both research- and mission-focused agencies such as NSF, NIH, NIST, DARPA, NOAA, DOE's Office of Science, and the D0D military service laboratories (see http://www.nitrd.gov/SUBCOMMITTEE/nitrd_agencies/index.aspx). There is research coordination between NITRD and Federal agencies conducting or supporting corresponding classified research.

[133] *Designing a Digital Future: Federally Funded Research and Development in Networking and Information Technology* (http://www.whitehouse.gov/sites/default/files/microsites/ostp/pcast-nitrd2013.pdf [2012] and http://www.whitehouse.gov/sites/default/files/microsites/ostp/pcast-nitrd-report-2010.pdf [2010]).

[134] Federal Networking and Information Technology Research and Development Program, "Report on Privacy Research Within NITRD [Networking and Information Technology Research and Development], National Coordination Office for NITRD, April 23, 2014. http://www.nitrd.gov/Pubs/Report_on_Privacy_Research_within_NITRD.pdf

[135] The Secure and Trustworthy Cyberspace program is the largest funder of relevant research. See: http://www.nsf.gov/funding/pgm_summ.jsp?pims_id=504709

[136] In December 2013, the NSF directorates supporting computer and social science joined in soliciting proposals for privacy-related research. http://www.nsf.gov/pubs/2014/nsf14021/nsf14021.jsp.

[137] See: http://www.iarpa.gov/index.php/research-programs/spar

[138] NIST is responsible for advancing the National Strategy for Trusted Identities in Cyberspace (NSTIC), which is intended to facilitate secure transactions within and across public and private sectors. See: http://www.nist.gov/nstic/

[139] Pike, W.A. et al., "PNNL [Pacific Northwest National Laboratory] Response to OSTP Big Data RFI," March 2014.

[140] A basis can be found in the newest version of the curriculum guidance of the Association for Computing Machinery (http://www.acm.org/education/CS2013-final-report.pdf). Given all of the pressures on curriculum, progress—as with cybersecurity—may hinge on growth in privacy-related research, business opportunities, and occupations.

[141] A beginning can be found in the Federal Government's FedRAMP program for certifying cloud services. Initiated to address Federal agency security concerns, FedRAMP already builds in attention to privacy in the form of a required Privacy Threshold Analysis and in some situations a Privacy Impact Analysis. The office of the U.S. Chief Information Officer provides guidance on Federal uses of information technology that addresses privacy along with security (see http://cloud.cio.gov/). It provides specific guidance on the cloud and FedRAMP (http://cloud.cio.gov/fedramp), including privacy protection (http://cloud.cio.gov/document/privacy-threshold-analysis-and-privacy-impact-assessment).

INDEX

#

20th century, 16, 55, 94
21st century, 67
9/11, 28

A

abstraction, 114
abuse, 7, 11, 36, 38, 67, 72
accessibility, 15, 71
accountability, 9, 23, 47, 51, 69, 71, 137, 146, 156
adjustment, 117
administrators, 29, 30
adolescents, 66
adults, 26, 66
advancement(s), 3, 67, 70
advertisements, 49
advocacy, 157
Afghanistan, 7
African-American, 55
age, 19, 27, 33, 34, 49, 62, 69, 108, 110, 119, 140
agencies, 4, 13, 14, 15, 19, 28, 31, 36, 37, 38, 39, 41, 44, 45, 46, 51, 58, 62, 66, 68, 69, 70, 71, 72, 73, 84, 86, 89, 146, 156, 161, 162, 168
aggregation, 53, 151
agriculture, 60

Air Force, 58, 156
air quality, 123
air temperature, 124
algorithm, 48, 94, 111, 126, 132
allergens, 115
alters, 50
American Civil Liberties Union, 74, 77, 81, 89, 144, 162
American History, 161
analog sources, vii, 93, 101, 124
applied mathematics, 126
aptitude, 27
arrest(s), 8, 37
articulation, 137, 149
artificial intelligence, 126
Asia, 18, 22, 30, 65, 87, 88
Asia Pacific Economic Cooperation, 18, 65
Asian Americans, 74
assessment, 25, 39, 80, 98, 154, 169
assets, 10, 12, 14, 91, 101, 162
asymmetry, 3, 40
attitudes, 4, 84, 118, 125
audit, 30, 70, 137, 146, 147, 153, 157, 167
authentication, 135, 136, 166
authenticity, 138
authority(s), 23, 24, 28, 29, 31, 33, 34, 51, 73, 106, 138, 139, 143, 152, 167
automate, 145, 167
autonomy, 11, 50, 110
awareness, 9, 32, 147

B

baby boomers, 113
banking, 22
bankruptcies, 45
banks, 34, 55
barometric pressure, 123
barriers, 39, 50, 54, 71, 108
base, 145, 148
basic research, 58, 72
beer, 117
behaviors, 8, 105, 127
benefits, vii, 8, 25, 27, 32, 36, 40, 41, 46, 52, 56, 61, 63, 65, 94, 108, 111, 134, 141, 151, 152
benign, 117
biological samples, 124
blogs, 86, 163
blueprint, 13
borrowers, 15
brainstorming, 14
brand loyalty, 117
Brazil, 167
browser, 43, 121
browsing, 5
budding, 158
building blocks, 96
bullying, 110
business model, 47, 67, 133
businesses, vii, 15, 17, 35, 39, 41, 44, 47, 49, 52, 60, 64, 66, 91, 121, 134

C

calculus, 1
campaigns, 41, 131
cancer, 59, 165
carbon, 40, 115
carbon monoxide, 40, 115
caricature, 137
case law, 16, 104
causality, 127
causation, 8, 127
cell phones, 112, 123, 124, 125, 134
Census, 12, 23, 74, 86, 162
certificate, 139, 140
certification, 22, 87
challenges, vii, 7, 9, 11, 12, 22, 26, 39, 57, 63, 65, 73, 86, 100, 103, 106, 111, 112, 114, 117, 122, 124, 128, 134, 137, 153
Chamber of Commerce, 83
chaos, 85
checks and balances, 106
chemical, 123, 124
Chicago, 32, 39, 89
childhood, 114
children, 19, 24, 26, 27, 42, 66, 110, 114, 118, 127, 128, 129
cholera, 1
cigarette smoke, 118
circulation, 104
city(s), vii, 30, 32, 39, 46, 53, 60, 86, 89, 123
citizens, 13, 14, 22, 23, 28, 38, 50, 51, 52, 54, 71, 87
citizenship, 29
civil law, 51
civil liberties, 28, 29, 30, 31, 38, 52, 68, 69, 70, 71, 72, 73
civil rights, 4, 23, 31, 48, 52, 54, 62, 68, 69, 70, 105
civil society, 73
clarity, 136
classes, 38, 48, 49, 53, 56, 62, 68, 101, 109, 110, 114, 127, 139, 151, 154, 155
classroom, 25, 26, 66, 152
climate, 13, 16, 38
climate change, 38
clinical symptoms, 24
clinical trials, 113
clustering, 131
clusters, 1, 8
codes, 21
codes of conduct, 21
Cold War, 50
collaboration, 58, 65, 70
collateral, 85
college students, 112
collisions, 105

Index

collusion, 140
color, 114, 125
commerce, 21, 47, 60, 94, 98, 99, 103, 131, 153, 154, 155
commercial, 6, 26, 34, 36, 37, 41, 44, 52, 70, 81, 98, 107, 109, 120, 127, 129, 132, 133, 140, 144, 146, 147, 149, 150, 152, 158, 162, 165
commodity, 27, 42
common law, 12, 16, 19, 103
communication, vii, 35, 58, 93, 101, 108, 116, 135, 145, 162, 166, 168
communication technologies, vii, 93, 101
Communications Act, 51
community(s), 10, 11, 26, 29, 38, 44, 48, 55, 66, 69, 70, 73, 95, 97, 98, 132, 133, 145, 146, 153, 157, 158, 161
competition, 48, 165
competitive advantage, 106
competitors, 106, 117
complement, 156
complexity, 6, 21, 24, 148
compliance, 25, 45, 58, 137, 147, 167
computation, 2, 140, 146, 157
computer, 3, 5, 6, 15, 33, 38, 48, 72, 95, 98, 101, 115, 116, 119, 120, 121, 125, 126, 127, 129, 135, 136, 137, 138, 139, 143, 145, 153, 154, 156, 157, 163, 165, 168
computer systems, 6, 136
computer use, 135
computing, vii, 5, 12, 17, 33, 51, 58, 93, 101, 103, 120, 158, 161, 166, 167, 168
conference, 166
confidentiality, 13, 19, 23, 71, 135, 136, 139
configuration, 164
conflict, 103
conformity, 153
confounding variables, 127
Congress, 10, 19, 34, 51, 62, 64, 69, 103, 162
consensus, 19, 43, 107
consent, 18, 27, 35, 44, 46, 50, 56, 57, 58, 59, 60, 63, 96, 97, 113, 114, 116, 122, 141, 144, 148, 150, 151, 152
Constitution, 10, 12, 16, 23, 33, 76, 87, 103

consulting, 141, 146
consumer advocates, 100
consumer protection, 62, 68, 69, 143
consumers, vii, 8, 9, 11, 13, 14, 19, 20, 21, 39, 40, 41, 42, 43, 44, 45, 46, 47, 48, 52, 53, 55, 57, 64, 67, 68, 72, 87, 109, 117, 148, 149, 150
consumption, 38, 56
controversial, 103
convergence, 95, 124, 126
conversations, 9, 65, 113, 139
cooperation, 10, 50
coordination, 71, 99, 156, 168
correlation(s), 8, 95, 105, 127, 128, 131
cost, 2, 5, 16, 24, 32, 45, 49, 56, 65, 101, 127, 142
Council of Europe, 18, 21
counseling, 114
course content, 25
Court of Appeals, 34
creativity, 39
credit history, 46
creditors, 46
creditworthiness, 40, 45, 46
crimes, 31, 32, 33, 131
criminal activity, 32, 33
criminal investigations, 69
criminal justice system, 69
criminals, 31, 32
crises, 163
critical infrastructure, 10, 13, 71
crop, 1
cryptography, 137, 138, 139, 156, 157, 167
CT, 123
culture, 39
cure, 59
curriculum, 67, 169
customer data, 128
customers, 15, 19, 40, 41, 45, 46, 55, 85, 111, 112, 117, 121, 128, 131, 145, 158
cybersecurity, 12, 40, 70, 71, 86, 96, 136, 156, 157, 166, 167, 169
cyberspace, 115
Czech Republic, 166

D

damages, 53
data analysis, 6, 8, 69, 70, 90, 149
data analytics, vii, 3, 5, 7, 9, 11, 12, 23, 24, 38, 39, 52, 62, 68, 71, 72, 81, 95, 96, 97, 98, 108, 109, 132, 133, 134, 150, 151, 153
data center, 49, 132, 133, 134
data collection, 1, 5, 42, 50, 53, 60, 96, 97, 99, 102, 109, 133, 149, 151, 155
data fusion, vii, 5, 9, 95, 98, 122, 128, 129, 131, 149, 151, 152, 153, 155, 165
data mining, vii, 55, 68, 94, 122, 126, 127, 128, 153, 157
data processing, 5, 95
data set, 8, 13, 29, 44, 56, 60, 71, 94, 97, 101, 120, 123, 126, 128, 150, 152, 155
database, 29, 36, 54, 113, 120, 131, 133, 140, 146, 164
defamation, 107
deficiencies, 103
de-identification capabilities, viii
democracy, 12, 59
democratization, 134
demonstrations, 86
denial, 19, 45
Department of Commerce, 62, 63, 64, 65, 75
Department of Defense, 37, 70, 90, 112, 156
Department of Education, 15, 26, 27, 79, 88, 89, 163
Department of Energy, 16, 38, 58
Department of Health and Human Services, 19, 87
Department of Homeland Security (DHS), 28, 29, 30, 54, 58, 70, 75, 89, 92
Department of Justice, 68, 89
Department of the Treasury, 36
destruction, 142, 143, 147
detectable, 8, 119
detection, 40, 45, 89, 124, 126, 129
deterrence, 98, 147, 150
diet, 23
digital communication, 51, 94
digital evidence, 32
dignity, 12, 86
dimensionality, 128
direct controls, 153
disclosure, 17, 19, 34, 35, 87, 94, 102, 107, 108, 122, 124, 150, 151, 152, 155, 162, 166
discrimination, 8, 11, 47, 50, 53, 54, 55, 61, 68, 94, 102, 104, 127, 162
diseases, 7, 24
dissenting opinion, 104
distribution, 63, 138
diversity, 96, 115, 120, 136, 142
DNA, 104, 124
doctors, 7, 24
draft, 62, 64
drawing, 4, 13, 32, 103
drug abuse, 38
Drug Enforcement Administration, 75

E

early warning, 7
eBay, 134
economic activity, 1
economic growth, 39, 96
economic incentives, 121
economics, 127
ecosystem, 24, 41, 42, 53, 134
ECPA, 51, 62, 69
education, vii, 13, 22, 25, 26, 27, 48, 49, 53, 55, 60, 61, 66, 67, 69, 87, 89, 99, 110, 114, 157, 169
educational institutions, 27, 66, 99, 157
educational research, 114
elaboration, 87
electricity, 7, 15, 40
electromagnetic, 123
electronic communications, 34, 51, 88
Electronic Communications Privacy Act, 34, 51, 62, 69, 88, 90
emotional state, 112, 119, 125
employees, 37, 38, 54, 55
employers, 54
employment, 44, 45, 53, 54, 55, 57, 67, 69

Index

encoding, 47
encouragement, 99, 147, 156, 157
encryption, 12, 58, 72, 96, 108, 135, 137, 138, 139, 140, 153, 156, 167
energy, vii, 5, 7, 13, 15, 38, 40, 49, 60, 87, 156
energy efficiency, 7, 15, 40
energy efficient, vii
enforcement, 20, 21, 22, 30, 31, 32, 34, 51, 70, 89, 129, 146
engineering, 122, 156
enrollment, 15
entrepreneurs, 13, 49, 87
environment(s), 5, 11, 12, 20, 63, 95, 114, 120, 124, 129, 130, 134, 143, 151, 152
epidemic, 32
Equal Credit Opportunity Act, 45, 47
Equal Employment Opportunity Commission, 68
equality, 55
equipment, 32, 40, 60
equity, 50
erosion, 131
ethical issues, 13
ethics, 86, 157
Europe, 18, 21, 65, 88
European Commission, 75, 87, 88
European Court of Justice, 88
European Union (EU), 18, 21, 22, 44, 65, 75, 87
everyday life, 119
evolution, 21, 25, 107, 153, 168
Executive Order, 13, 14, 86
exercise, 4, 20, 23, 148, 149
expertise, 62, 68, 70, 99, 146, 157
extraction, 129, 130, 165

F

Facebook, 75, 91, 130, 132, 134, 163, 166, 167
facial expression, 117, 122
factories, vii
fairness, 11, 19, 37, 50, 55, 61, 68, 112, 127
false alarms, 116

false negative, 105
false positive, 105
families, 15, 67
fantasy, 97, 141
FDIC, 91
Federal Bureau of Investigation (FBI), 17, 75, 165
federal courts, 34
federal government, 11, 12, 13, 14, 18, 19, 34, 38, 39, 49, 51, 62, 67, 68, 69, 70, 73, 90, 169
federal law, 30, 65, 70, 105
Federal Register, 73
Federal Student Aid, 15
Federal Trade Commission Act, 24
feelings, 84
fiber, 132
fidelity, 10, 25, 106, 128
films, 8
filters, 48
financial, 3, 19, 22, 24, 37, 42, 48, 52, 53, 73, 115, 117, 120, 140, 146, 154, 163
financial crisis, 117
financial data, 24, 42
financial innovation, 163
financial institutions, 19
financial records, 115, 117
financial sector, 19, 140
First Amendment, 37, 51, 162
first generation, 26
food, 22, 115, 117
football, 117, 132
force, 49, 106, 118, 119
Ford, 91
foreign nationals, 28
formal language, 137
Fort Hood, 37
foundations, 58
Fourth Amendment, 12, 16, 33, 34, 86, 87, 103, 108, 115
framing, 22, 85
fraud, 7, 22, 36, 38, 40, 43, 44, 45, 72, 85, 110
free association, 51
free goods, 52

fruits, 98, 150, 154
FSB, 137
funding, 81, 99, 156, 168
fusion, vii, 5, 8, 9, 95, 98, 122, 128, 129, 131, 136, 149, 151, 152, 153, 155, 165

G

gait, 113, 119, 123
gambling, 112, 162
gamma radiation, 123
GAO, 91
genes, 24
genetic information, 19, 24
genome, 94, 102
geography, 108, 129
geolocation, 129
gestures, 116, 123, 163
global economy, 39
goods and services, 48, 49, 53, 67, 68, 109
google, 91, 163, 165
Google Earth, 129
government procurement, 158
governments, 11, 24, 49, 106
GPS, 5, 6, 31, 53, 89, 94, 120, 129
grades, 114
Gramm-Leach-Bliley Act, 19, 24
grants, 14, 39, 70, 89, 156
graph, 163, 165, 166
grids, 7, 40
growth, vii, 39, 41, 93, 101, 120, 166, 169
guardian, 27
guidance, 11, 26, 27, 36, 37, 163, 169
guidelines, 18, 21, 36

H

Harvard Law School, 159, 160
Hawaii, 92
health, vii, 3, 7, 13, 14, 18, 19, 22, 23, 24, 25, 27, 38, 44, 47, 49, 55, 58, 60, 61, 65, 67, 69, 73, 85, 87, 88, 94, 104, 109, 112, 113, 115, 117, 123, 125, 127, 140, 142, 152, 155, 156

health care, 7, 14, 18, 23, 24, 25, 27, 44, 49, 53, 58, 60, 65, 69, 73, 85, 87, 88, 127
health care system, 23
health information, 14, 19, 24, 25, 65, 87, 155
health insurance, 115
health status, 24, 125
heart rate, 5, 7, 112
height, 50
heroin, 116
heterogeneity, 135
hiring, 39
history, 10, 12, 13, 33, 37, 41, 46, 55, 86, 94, 117, 129, 146
homeland security, 22, 29, 69, 70, 129
homeowners, 117
homes, vii, 15, 60, 103, 115
homework, 25
host, 47
hotel, 118
hotspots, 32
House, 4, 12, 162
House of Representatives, 12, 162
household income, 128
housing, 47, 53, 55, 69, 128
human, 2, 16, 18, 30, 31, 124, 129, 130, 136, 142, 145, 146, 158, 166
human brain, 166
human genome, 16
human right, 18
human subjects, 142

I

icon, 42
ID, 121
identification, viii, 6, 9, 56, 72, 86, 88, 95, 105, 110, 115, 120, 122, 123, 129, 130, 131, 138, 142, 149, 152, 167
identity, 3, 9, 11, 24, 26, 28, 32, 44, 45, 47, 52, 54, 55, 85, 105, 119, 125, 135, 136, 138, 139, 156, 162, 165, 166, 168
illumination, 122
image(s), 10, 124, 125, 126, 128, 129, 134, 144, 162

imagery, 129, 165
imagination, 39
immunization, 15
impact assessment, 30, 89
improvements, 30, 52, 65, 119
income, 48, 127
individual action, 64
individual privacy, vii, 9, 11, 12, 36, 69, 80, 93, 94, 100, 101, 106, 115
individual rights, 144
individual students, 114
industrial economy, vii
industrial revolution, 16
industry(s), 1, 4, 18, 24, 41, 42, 44, 47, 49, 52, 64, 73, 88, 91, 100, 112, 116, 146, 153, 157, 161
infancy, 127
infection, 7
inferences, 94, 95, 102, 108, 112, 119, 128
information exchange, 134
information processing, 86, 101
information sharing, 70, 71
information technology, 25, 101, 169
infrastructure, 40, 69, 72, 103, 108, 116, 119, 132, 133, 134, 147, 153
ingest, 24
inhibitor, 137
inspectors, 54
institutions, 11, 15, 50, 61, 66
integration, 49, 58, 128
integrity, 20, 98, 135, 154
intelligence, 4, 11, 28, 29, 31, 36, 37, 60, 69, 70, 73, 84, 85, 86, 128, 132, 146, 161
intelligent systems, 128
intensive care unit, 7
interface, 115
interference, 22, 64, 93, 102, 104
intermediaries, 97
Internal Revenue Service, 15
international law, 81
international trade, 125
interoperability, 22, 25, 128
intervention, 31, 141
intimacy, 103
invasion of privacy, 17
inventions, 16, 85, 139
investment(s), 3, 12, 39, 40, 56, 67, 72, 157
IP address, 35, 132
Ireland, 88
isolation, 105, 108, 125
issues, 4, 11, 13, 17, 24, 27, 35, 38, 51, 54, 81, 83, 94, 97, 102, 105, 106, 114, 124, 127, 145, 153, 161, 166
iteration, 66

J

Jews, 55
Jordan, 158
journalists, 101
justification, 17

K

kill, 67

L

landscape, 11, 50, 80
lasers, 129
Latinos, 55
law enforcement, 4, 22, 29, 30, 31, 32, 33, 36, 51, 61, 64, 69, 70, 73, 81, 84, 89, 106, 110
laws, 8, 11, 12, 16, 18, 19, 21, 22, 24, 31, 35, 51, 58, 68, 87, 91, 94, 98, 142, 147, 150, 163
laws and regulations, 22
lawyers, 17, 142
lead, 5, 9, 11, 33, 38, 52, 62, 64, 65, 68, 72, 94, 99, 109, 113, 114, 124, 128, 145, 157
leadership, 13, 39, 100, 158
learning, 23, 25, 26, 27, 48, 61, 66, 67, 110, 111, 112, 114, 126, 127, 140, 162
learning disabilities, 27
learning outcomes, 26, 114
learning process, 25
learning styles, 110, 114
LED, 119, 165

legal protection, 51, 65, 94, 107
legislation, 20, 23, 53, 62, 64, 71, 105, 152, 154
leisure, vii
leisure time, vii
liberty, 90
lidar, 165
life cycle, 149
light, 17, 22, 34, 35, 69, 103, 107, 116
literacy, 61, 67
litigation, 32
loans, 13, 15
local government, 27
location information, 35, 90, 122, 151
locus, 98, 115, 133, 152, 154
logging, 147
logistics, 40
love, 115
low risk, 126
lying, 140

M

machine learning, 101, 126, 127, 162, 165
magazines, 117
magnetic field(s), 123
magnitude, 125
majority, 29, 33, 54, 57, 87, 104, 114, 161
man, 86
management, 2, 5, 33, 36, 40, 60, 72, 85, 87, 99, 114, 120, 128, 146, 156, 157, 162, 168
manufacturing, 49
mapping, 5, 6, 22, 129
marijuana, 115
market access, 53
market failure, 97, 141
market penetration, 146
market position, 56
market share, 145
marketing, 41, 44, 45, 47, 52, 53, 64, 88, 117, 121, 131, 132, 166
marketplace, 8, 17, 39, 41, 42, 57, 61, 96, 97, 106, 133, 135, 145
marriage, 1

Maryland, 34, 76, 90, 165
mass, 94
mass media, 94
materials, 25, 114
mathematics, 72
matter, 10, 27, 36, 45, 51, 53, 97, 116, 139, 151, 155
measurement(s), 3, 152
media, 5, 6, 26, 37, 38, 41, 42, 45, 47, 84, 87, 105, 112, 120, 130, 134, 138, 165, 166
Medicaid, 7, 85
medical, 15, 24, 42, 59, 65, 84, 94, 101, 113, 123, 142, 165
medical science, 65
Medicare, 7, 85
medicine, 24
memory, 38, 133
messages, 35, 138, 139, 164
methodology, 89
Mexico, 159
Microsoft, 76, 77, 78, 79, 80, 82, 83, 91, 116, 130, 134, 144, 146, 147
military, 7, 26, 86, 161, 168
minorities, 98, 108, 154
minors, 89, 155
mission(s), 28, 29, 30, 38, 70, 71, 161, 168
misuse, 26, 95, 102, 114, 120, 125, 136, 138
mobile device, 5, 6, 52, 113, 115, 120, 121, 134, 141, 144
mobile phone, 51, 52, 123, 166
models, 10, 24, 129, 134, 165
monopoly, 106
monopoly power, 106
Moon, 159
morbidity, 127
mosaic, 9
MPI, 133
MRI, 123
music, 2

N

National Academy of Sciences, 165, 166
National Economic Council, 73, 76

Index

national emergency, 23
National Institutes of Health, 16
National Research Council, 164
national security, 31, 34, 60, 61, 69, 70, 71, 106
National Security Agency (NSA), 37, 58, 76, 77, 133, 137, 156, 167
National Strategy, 169
National Survey, 91
nationality, 62, 66
natural hazards, 109
navigation system, 13, 129
negative consequences, 96, 104, 127
negotiating, 29, 125
negotiation, 25, 97, 141, 145
Netherlands, 139
networking, 103, 132, 133, 161, 168
neutral, 11, 48
next generation, 60, 133
Nidal Hasan, 37
NOAA, 161, 168
non-citizens, 54
non-repudiation, 135

O

Obama, 12, 13, 14, 21, 39, 58, 86, 100
Obama Administration, 12, 13, 14, 21, 39, 58
objectivity, 48
Office of Management and Budget (OMB), 13, 14, 36, 37, 62, 66, 86, 90
officials, 4, 14, 22, 30, 31, 71, 73, 92
online advertising, 41, 42, 44, 47, 52, 63, 64, 89
online learning, 26, 27, 155
openness, 13
operating system, 28
operations, 45, 68, 133
opportunities, vii, 6, 8, 9, 26, 27, 38, 39, 47, 53, 60, 61, 66, 71, 99, 110, 125, 157, 158, 163, 169
opt out, 42, 43
Organization for Economic Cooperation and Development (OECD), 18, 21, 65, 87, 88

organize, 16, 60
oversight, 28, 29, 30, 31, 69, 70, 84, 134
ownership, 27, 117

P

Pacific, 22, 83, 87, 88, 159, 169
parallel, 4, 5, 60, 80, 133
parental consent, 19
parents, 26, 118
participants, 53, 73
pathways, 25
pattern recognition, 122
payroll, 120
peace, 22
pedagogy, 114
penalties, 98, 147, 150
performers, 72
permission, 60, 63, 121
permit, 43, 45
personal computers, 51
personal control, 10
personal life, 107
pharmaceutical, 113, 146
Philadelphia, 32, 89
photographs, 104
physical activity, 5
physicians, 114
platform, 4, 14, 27, 73, 114, 133, 146
playing, 26, 60, 97, 141
police, 30, 31, 32, 33, 50, 90, 111, 116, 118
policy choice, 161
policy instruments, 49
policy options, 140
policymakers, 17, 22, 91
political participation, 108
pollutants, 115, 123
population, 1, 8, 37, 48, 68, 110, 117, 166
portfolio, 28
portraits, 104
power lines, 127
pragmatism, 92
precedents, 154
predicate, 33
preparation, 127

prescription drugs, 23
preservation, 157
president, v, 1, 2, 4, 5, 10, 11, 12, 13, 16, 25, 36, 56, 57, 58, 59, 60, 62, 63, 64, 66, 68, 73, 77, 78, 79, 80, 85, 86, 88, 90, 92, 93, 100, 161, 163
President Obama, 4, 10, 11, 12, 13, 16, 59, 63, 73
presumption of innocence, 69
prevention, 33, 85, 129
principles, 2, 18, 20, 21, 42, 63, 85, 91, 103, 136, 137, 147, 148, 149, 164
Privacy Protection Act, 19, 26, 27, 66, 67, 88
privacy-sensitive information, viii
private information, 94, 106, 109
private sector, vii, 10, 13, 14, 15, 17, 22, 39, 46, 50, 52, 53, 60, 65, 70, 73, 80, 81, 97, 100, 102, 106, 143, 154, 169
probability, 1, 48, 126, 147
probability theory, 1
procurement, 100, 158
professionals, 24, 99, 145, 157
profit, 87
programming, 89, 131, 133, 135, 147
project, 7, 22, 39, 92, 161, 165
proliferation, 31
property rights, 117
proposition, 9
prosperity, 2, 12, 71
public awareness, 69
public concern, 107
public goods, 119
public health, 23, 127
public interest, 50
public opinion, 105
public policy, 65, 81
public resources, 33
public safety, 13, 31, 40, 51, 61, 70, 119, 123
public sector, 15, 22, 23, 38, 39, 49, 70, 147
public service, 38, 39, 50, 52, 60, 61, 71
publishing, 107, 161, 162

Q

quality of life, 3, 12
Quartz, 163
query, 7, 113, 140
questionnaire, 152

R

race, 45, 94, 102
racial minorities, 109
radar, 124
radiation, 123
radio, 6, 94, 111, 115
RE, 167
reading, 114, 116, 117, 125, 130, 144
real time, 5, 7, 8, 125
reality, 56, 140
reasoning, 146
recognition, 6, 56, 105, 107, 116, 117, 118, 119, 125, 129, 130, 151, 165
recommendations, 36, 40, 62, 91, 98, 101, 103, 154, 156, 161
recruiting, 134
reform(s), 38, 65, 88, 162
registries, 15
regulations, 19, 62, 65, 81, 98, 137, 147, 150, 157
regulatory framework, 41, 42, 66, 67
relevance, 115
reliability, 71, 115, 146
religion, 45
religious beliefs, 162
remote sensing, 101
rent, 40, 117
reproduction, 104
reputation, 121
requirements, 14, 21, 22, 25, 27, 36, 43, 69, 88, 137, 145, 153
research funding, 16
researchers, 6, 8, 23, 25, 65, 96, 130, 132, 143, 167
resilience, 71
resolution, 124, 125, 128

Index

resources, 31, 50, 71, 72, 85, 86, 91, 97, 134, 143
respiration, 123
response, 10, 32, 56, 97, 128, 143
restrictions, 51, 133, 137
retail, 15
retaliation, 162
rights, 10, 13, 14, 16, 17, 19, 20, 22, 33, 34, 37, 38, 45, 51, 55, 63, 68, 69, 85, 86, 88, 89, 91, 102, 103, 104, 106, 107, 108, 115, 117, 149, 165
rings, 30
risk(s), vii, 7, 18, 20, 22, 24, 29, 36, 37, 39, 45, 63, 96, 106, 112, 113, 134, 136, 140, 143, 148, 150, 153
robotics, 127, 128
ROI, 166
root, 7, 41
routes, 109
rule of law, 35
rules, 18, 21, 28, 29, 31, 45, 51, 106, 143, 157

S

safety, 6, 38, 40, 66, 115, 116, 117
scaling, 133
schizophrenia, 8
school, 27, 40, 61, 67, 127, 155, 163
science, 1, 2, 12, 38, 58, 72, 86, 99, 110, 114, 126, 127, 156, 157, 161, 168
scientific knowledge, 72
scope, 3, 9, 47, 53, 88, 90, 103, 136
Second World, 23
Secretary of Commerce, 4, 73, 77, 78
Secretary of Defense, 90
security, 9, 10, 12, 13, 19, 20, 28, 31, 37, 38, 41, 43, 47, 52, 58, 69, 73, 86, 90, 96, 99, 100, 115, 116, 118, 119, 123, 125, 132, 134, 135, 136, 137, 148, 153, 156, 157, 158, 165, 166, 169
security practices, 20, 38, 148
security services, 132
seeding, 126
Senate, 91

sensations, 86
senses, 94, 119
sensing, 124, 129
sensitivity, 20, 37, 47, 148
sensor network, 128
sensors, 3, 5, 6, 32, 56, 105, 115, 117, 120, 123, 124, 125, 128, 134, 151, 165
servers, 43, 132, 134, 144, 167
service provider, 34, 158
sex, 30
sexuality, 104
shape, 60
short supply, 39
signals, 4, 11, 35, 60, 73, 86, 125, 161
signs, 5, 7, 112, 119
silver, 137, 167
small businesses, 132, 158
small firms, 40
smoking, 116
snaps, 144
social benefits, 151
social influence, 47, 52
social network, 32, 108, 121, 125, 130, 132, 165, 166
social norms, 3, 145
Social Security, 54
Social Security Administration, 54
society, 1, 12, 17, 48, 50, 52, 55, 60, 94, 102, 155, 162
software, 3, 7, 14, 16, 26, 32, 52, 66, 99, 116, 117, 122, 132, 133, 137, 145, 146, 150, 151, 157, 161, 166, 167, 168
solitude, 17, 107
solution, 43, 44, 103
specifications, 137
specter, 27, 48
speech, 4, 33, 51, 69, 100, 104, 129, 130, 151, 161, 165
spending, 36
stakeholders, 21, 62, 64, 65, 73
standardization, 96
stars, 1
state(s), 6, 12, 13, 15, 17, 19, 20, 24, 26, 27, 32, 34, 39, 53, 64, 70, 80, 87, 100, 104,

105, 106, 107, 125, 130, 147, 149, 154, 157, 165
state laws, 64, 154
state legislatures, 34
statistical inference, 152
statistics, 1, 126, 128, 140
statutes, 34
stock, 65, 112
storage, 2, 5, 12, 21, 32, 33, 38, 53, 56, 57, 70, 81, 97, 99, 101, 108, 115, 139, 143, 145, 148, 155
structure, 11, 131
subscribers, 116
substitutes, 115, 165
suicide, 1
supervision, 26, 33
supervisors, 51
suppliers, 15
supply chain, 60
support staff, 134
Supreme Court, 16, 31, 33, 34, 87, 90, 104, 109
surveillance, 7, 31, 32, 33, 50, 70, 120, 123, 125, 129, 130
sustainability, 87
Switzerland, 18, 87
Sybil, 166
synthesis, 146

T

talent, 3, 39, 72
target, 43, 49, 112, 131, 153, 167
taxonomy, 29, 59
taxpayers, 15
teachers, 25, 26
teams, 7
technical assistance, 39, 70
techniques, 8, 9, 16, 31, 33, 40, 58, 67, 70, 96, 101, 114, 121, 122, 124, 126, 128, 130, 131, 135, 138, 139, 140, 142, 149, 156
technological ability, vii, 3
technological advancement, 73
technological advances, 11, 34

technological revolution, vii
telecommunications, 35, 103
telephone, 17, 34, 103, 108, 120
telephone companies, 34
telephone numbers, 34
temperature, 7, 40
tenants, 55, 117
tension(s), 33, 79, 103
terrorists, 6, 127
test scores, 127
testing, 114
text messaging, 51, 120
theft, 26, 47, 120, 139
thoughts, 86, 112
threats, 40, 71, 94, 106, 137
time periods, 64
tobacco, 115
top-down, 18
tracks, 5, 114
trade, 21, 127, 163
traditional practices, 7
traditions, 65
trafficking, 30
training, 99, 134, 157
traits, 113
trajectory, 2, 10, 25, 43, 57
transactions, 3, 6, 15, 35, 41, 52, 120, 125, 126, 169
transformation, 52
transmission, 35, 128, 164
transparency, 14, 21, 42, 43, 47, 54, 55, 64, 68, 84, 89, 111, 143, 153, 156
transportation, 40, 95
Treasury, 36
treatment, 11, 23, 24, 40, 55, 60, 94, 113, 141
triangulation, 6
triggers, 38
tuition, 40

U

U.S. Department of Commerce, 87
U.S. economy, 41
U.S. policy, 81

Index 181

UK, 76, 112, 162
ultrasound, 116
UN, 163
United States, 2, 4, 10, 12, 13, 16, 17, 18, 19, 21, 22, 26, 28, 30, 32, 33, 34, 41, 54, 55, 61, 65, 66, 86, 87, 89, 90, 99, 102, 104, 133, 146, 157, 161, 162, 167
universe, 52
universities, 81, 87, 101
unreasonable searches, 12
updating, 90, 96
urban, 72, 129, 164

V

variables, 8, 33, 127
varieties, 22
vehicles, 2, 129
vein, 145
velocity, vii, 3, 5, 101
venture capital, 3
vibration, 123
victims, 47
video games, 110, 123
videos, 25, 105, 115, 129, 152
violence, 37
violent crime, 32
vision, 114, 116, 117, 130, 166
visualization, 129, 131
vulnerable people, 30

W

war, 7, 23
War on Terror, 90

Washington, 37, 75, 90, 159, 162, 163
waste, 7, 36, 38, 72
watches, 114, 117
water, 22
wealth, 30, 52, 147
weapons, 31
weapons of mass destruction, 31
weather patterns, 38
web, 5, 6, 8, 30, 41, 43, 45, 52, 68, 83, 86, 97, 112, 126, 127, 139, 141, 164, 167
web browser, 43, 139
web pages, 5
web service, 97, 141
webpages, 43
websites, 30, 42, 43, 44
well-being, 2, 12
White House, 4, 19, 59, 60, 73, 81, 83, 84, 85, 86, 87, 88, 92, 161, 163
White Paper, 167
Wi-Fi, 116
wireless networks, 3, 6
wiretaps, 16
workers, 26, 54
workforce, 48
World Wide Web, 44, 165, 166
worldwide, 2, 66
wrestling, 38

Y

Yale University, 86, 88, 164
yield, 38, 99, 105, 125, 127, 128, 151, 155
young adults, 66
young people, 27